# Making Friends

"Food will be brought," announced the Callean. "You should make your needs known; you are of low intelligence and helpless. I forbid nothing, I know you are harmless, and your life is short in any case; but I do not want you to get in the way."

"Thank you," Oberholzer said, and bringing the Sussmann gun into line, he trained it on the Callean's squidlike head and pulled the trigger.

*Then Oberholzer grounded the rifle and waited to see what would happen next.*

# Other SIGNET Science Fiction Adventures

# GALACTIC CLUSTER

by
## JAMES BLISH

A SIGNET BOOK from
NEW AMERICAN LIBRARY
TIMES MIRROR

SIXTH PRINTING

 SIGNET TRADEMARK REG. U.S. PAT. OFF. AND FOREIGN COUNTRIES
REGISTERED TRADEMARK—MARCA REGISTRADA
HECHO EN CHICAGO, U.S.A.

SIGNET, SIGNET CLASSICS, SIGNETTE, MENTOR AND PLUME BOOKS
*are published by The New American Library, Inc.,*
*1301 Avenue of the Americas, New York, New York 10019*

PRINTED IN THE UNITED STATES OF AMERICA

# Contents

TOMB TAPPER    7

KING OF THE HILL    27

COMMON TIME    38

A WORK OF ART    59

TO PAY THE PIPER    74

NOR IRON BARS    88

BEEP    120

THIS EARTH OF HOURS    156

TO KENNETH S. WHITE

# Tomb Tapper

THE DISTANT glare of the atomic explosion had already faded from the sky as McDonough's car whirred away from the blacked-out town of Port Jervis and turned north. He was making fifty m.p.h. on U.S. Route 209 using no lights but his parkers, and if a deer should bolt across the road ahead of him he would never see it until the impact. It was hard enough to see the road.

But he was thinking, not for the first time, of the old joke about the man who tapped train wheels.

He had been doing it, so the story ran, for thirty years. On every working day he would go up and down both sides of every locomotive that pulled into the yards and hit the wheels with a hammer; first the drivers, then the trucks. Each time, he would cock his head, as though listening for something in the sound. On the day of his retirement, he was given a magnificent dinner, as befitted a man with long seniority in the Brotherhood of Railway Trainmen—and somebody stopped to ask him what he had been tapping for all those years.

He had cocked his head as though listening for something, but evidently nothing came. "I don't know," he said.

That's me, McDonough thought. I tap tombs, not trains. But what am I listening for?

The speedometer said he was close to the turnoff for the airport, and he pulled the dimmers on. There it was. There was at first nothing to be seen, as the headlights swept along the dirt road, but a wall of darkness deep as all night, faintly edged at the east by the low domed hills of the Neversink valley. Then another pair of lights snapped on behind him, on the main highway, and came jolting after McDonough's car, clear and sharp in the dust clouds he had raised.

He swung the car to a stop beside the airport fence and killed the lights; the other car followed. In the renewed blackness the faint traces of dawn on the hills were wiped out, as though the whole universe had been set back an hour. Then the yellow eye of a flashlight opened in the window of the other car and stared into his face.

7

He opened the door. "Martinson?" he said tentatively.

"Right here," the adjutant's voice said. The flashlight's oval spoor swung to the ground. "Anybody else with you?"

"No. You?"

"No. Go ahead and get your equipment out. I'll open up the shack."

The oval spot of light bobbed across the parking area and came to uneasy rest on the combination padlock which held the door of the operations shack secure. McDonough flipped the dome light of his car on long enough to locate the canvas sling which held the components of his electro-encephalograph, and eased the sling out onto the sand.

He had just slammed the car door and taken up the burden when little chinks of light sprang into being in the blind windows of the shack. At the same time, cars came droning out onto the field from the opposite side, four of them, each with its wide-spaced unblinking slits of paired parking lights, and ranked themselves on either side of the landing strip. It would be dawn before long, but if the planes were ready to go before dawn, the cars could light the strip with their brights.

We're fast, McDonough thought, with brief pride. Even the Air Force thinks the Civil Air Patrol is just a bunch of amateurs, but we can put a mission in the air ahead of any other CAP squadron in this county. We can scramble.

He was getting his night vision back now, and a quick glance showed him that the windsock was flowing straight out above the black, silent hangar against the pearly false dawn. Aloft, the stars were paling without any cloud-dimming, or even much twinkling. The wind was steady north up the valley; ideal flying weather.

Small lumpy figures were running across the field from the parked cars toward the shack. The squadron was scrambling.

"Mac!" Martinson shouted from inside the shack. "Where are you? Get your junk in here and get started!"

McDonough slipped inside the door, and swung his EEG components onto the chart table. Light was pouring into the briefing room from the tiny office, dazzling after the long darkness. In the briefing room the radio blinked a tiny red eye, but the squadron's communications officer hadn't yet arrived to answer it. In the office, Martinson's voice rumbled softly, urgently, and the phone gave him back thin unintelligible noises, like an unteachable parakeet.

Then, suddenly, the adjutant appeared at the office door

and peered at McDonough. "What are you waiting for?" he said. "Get that mind reader of yours into the Cub on the double."

"What's wrong with the Aeronca? It's faster."

"Water in the gas; she ices up. We'll have to drain the tank. This is a hell of a time to argue." Martinson jerked open the squealing door which opened into the hangar, his hand groping for the light switch. McDonough followed him, supporting his sling with both hands, his elbows together. Nothing is quite so concentratedly heavy as an electronics chassis with a transformer mounted on it, and four of them make a back-wrenching load.

The adjutant was already hauling the servicing platform across the concrete floor to the cowling of the Piper Cub. "Get your stuff set," he said. "I'll fuel her up and check the oil."

"All right. Doesn't look like she needs much gas."

"Don't you ever stop talkin'? Let's move."

McDonough lowered his load to the cold floor beside the plane's cabin, feeling a brief flash of resentment. In daily life Martinson was a job printer who couldn't, and didn't, give orders to anybody, not even his wife. Well, those were usually the boys who let rank go to their heads, even in a volunteer outfit. He got to work.

Voices sounded from the shack, and then Andy Persons, the commanding officer, came bounding over the sill, followed by two sleepy-eyed cadets. "What's up?" he shouted. "That you, Martinson?"

"It's me. One of you cadets, pass me up that can. Andy, get the doors open, hey? There's a Russki bomber down north of us, somewhere near Howells. Part of a flight that was making a run on Schenectady."

"Did they get it?"

"No, they overshot, *way* over—took out Kingston instead. Stewart Field hit them just as they turned to regroup, and knocked this baby down on the first pass. We're supposed to——"

The rest of the adjutant's reply was lost in a growing, echoing roar, as though they were all standing underneath a vast trestle over which all the railroad trains in the world were crossing at once. The sixty-four-foot organ reeds of jets were being blown in the night zenith above the field—another hunting pack, come from Stewart Field to avenge the hydrogen agony that had been Kingston.

His head still inside the plane's greenhouse, McDonough

listened transfixed. Like most CAP officers, he was too old to be a jet pilot, his reflexes too slow, his eyesight too far over the line, his belly muscles too soft to take the five-gravity turns; but now and then he thought about what it might be like to ride one of those flying blowtorches, cruising at six hundred miles an hour before a thin black wake of kerosene fumes, or being followed along the ground at top speed by the double wave-front of the "supersonic bang." It was a noble notion, almost as fine as that of piloting the one-man Niagara of power that was a rocket fighter.

The noise grew until it seemed certain that the invisible jets were going to bullet directly through the hangar, and then dimmed gradually.

"The usual orders?" Persons shouted up from under the declining roar. "Find the plane, pump the live survivors, pick the corpses' brains? Who else is up?"

"Nobody," Martinson said, coming down from the ladder and hauling it clear of the plane. "Middletown squadron's deactivated; Montgomery hasn't got a plane; Newburgh hasn't got a field."

"Warwick has Group's L-16——"

"They snapped the undercarriage off it last week," Martinson said with gloomy satisfaction. "It's our baby, as usual. Mac, you got your ghoul-tools all set in there?"

"In a minute," McDonough said. He was already wearing the Walter goggles, pushed back up on his helmet, and the detector, amplifier, and power pack of the EEG were secure in their frames on the platform behind the Cub's rear seat. The "hair net"—the flexible network of electrodes which he would jam on the head of any dead man whose head had survived the bomber crash—was connected to them and hung in its clips under the seat, the leads strung to avoid fouling the plane's exposed control cables. Nothing remained to do now but to secure the frequency analyzer, which was the heaviest of the units and had to be bolted down just forward of the rear joystick so that its weight would not shift in flight. If the apparatus didn't have to be collimated after every flight, it could be left in the plane—but it did, and that was that.

"O.K.," he said, pulling his head out of the greenhouse. He was trembling slightly. These tomb-tapping expeditions were hard on the nerves. No matter how much training in the art of reading a dead mind you may have had, the actual experience is different, and cannot be duplicated from the long-stored corpses of the laboratory. The newly dead brain is an inferno, almost by definition.

"Good," Persons said. "Martinson, you'll pilot. Mac, keep on the air; we're going to refuel the Airoknocker and get it up by ten o'clock if we can. In any case we'll feed you any spottings we get from the Air Force as fast as they come in. Martinson, refuel at Montgomery if you have to; don't waste time coming back here. Got it?"

"Roger," Martinson said, scrambling into the front seat and buckling his safety belt. McDonough put his foot hastily into the stirrup and swung into the back seat.

"Cadets!" Persons said. "Pull chocks! Roll 'er!"

Characteristically, Persons himself did the heavy work of lifting and swinging the tail. The Cub bumped off the apron and out on the grass into the brightening morning.

"Switch off!" the cadet at the nose called. "Gas! Brakes!"

"Switch off, brakes," Martinson called back. "Mac, where to? Got any ideas?"

While McDonough thought about it, the cadet pulled the prop backwards through four turns. "Brakes! Contact!"

"Let's try up around the Otisville tunnel. If they were knocked down over Howells, they stood a good chance to wind up on the side of that mountain."

Martinson nodded and reached a gloved hand over his head. "Contact!" he shouted, and turned the switch. The cadet swung the prop, and the engine barked and roared; at McDonough's left, the duplicate throttle slid forward slightly as the pilot "caught" the engine. McDonough buttoned up the cabin, and then the plane began to roll toward the far, dim edge of the grassy field.

The sky got brighter. They were off again, to tap on another man's tomb, and ask of the dim voice inside it what memories it had left unspoken when it had died.

The Civil Air Patrol is, and has been since 1941, an auxiliary of the United States Air Force, active in coastal patrol and in air-sea rescue work. By 1954—when its ranks totaled more than eighty thousand men and women, about fifteen thousand of them licensed pilots—the Air Force had nerved itself up to designating CAP as its Air Intelligence arm, with the job of locating downed enemy planes and radioing back information of military importance.

Aerial search is primarily the task of planes which can fly low and slow. Air Intelligence requires speed, since the kind of tactical information an enemy wreck may offer can grow cold within a few hours. The CAP's planes, most of them single-engine, private-flying models, had already been proven ideal aerial search instruments; the CAP's radio net, with its more than seventy-five hundred fixed, mobile and airborne stations, was more than fast enough to get information to wherever it was needed while it was still hot.

But the expected enemy, after all, was Russia; and how many civilians, even those who know how to fly, navigate, or operate a radio transmitter, could ask anyone an intelligent question in Russian, let alone understand the answer?

It was the astonishingly rapid development of electrical methods for probing the brain which provided the answer—in particular the development, in the late fifties, of flicker-stimulus aimed at the visual memory. Abruptly, EEG technicians no longer needed to use language at all to probe the brain for visual images, and read them; they did not even need to know how their apparatus worked, let alone the brain. A few moments of flicker into the subject's eyes, on a frequency chosen from a table, and the images would come swarming into the operator's toposcope goggles—the frequency chosen without the slightest basic knowledge of electrophysiology, as a woman choosing an ingredient from a cookbook is ignorant of—and indifferent to—the chemistry involved in the choice.

It was that engineering discovery which put tomb-tappers into the back seats of the CAP's putt-putts when the war finally began—for the images in the toposcope goggles did not stop when the brain died.

The world at dawn, as McDonough saw it from three thousand feet, was a world of long sculptured shadows, almost as motionless and three-dimensional as a lunar landscape near the daylight terminator. The air was very quiet, and the Cub droned as gently through the blue haze as any bee, gaining altitude above the field in a series of wide climbing turns. At the last turn the plane wheeled south over a farm owned by someone Martinson knew, a man already turning his acres from the seat of his tractor, and Martinson waggled the plane's wings at him and got back a wave like the quivering of an insect's antenna. It was all deceptively normal.

Then the horizon dipped below the Cub's nose again and Martinson was climbing out of the valley. A lake passed below them, spotted with islands, and with the brown barracks of

Camp Cejwin, once a children's summer camp but now full of sleeping soldiers. Martinson continued south, skirting Port Jervis, until McDonough was able to pick up the main line of the Erie Railroad, going northeast toward Otisville and Howells. The mountain through which the Otisville tunnel ran was already visible as a smoky hulk to the far left of the dawn.

McDonough turned on the radio, which responded with a rhythmical sputtering; the Cub's engine was not adequately shielded. In the background, the C.O.'s voice was calling them: "Huguenot to L-4. Huguenot to L-4."

"L-4 here. We read you, Andy. We're heading toward Otisville. Smooth as glass up here. Nothing to report yet."

"We read you weak but clear. We're dumping the gas in the Airoknocker *crackle* ground. We'll follow as fast as possible. No new AF spottings yet. If *crackle*, call us right away. Over."

"L-4 to Huguenot. Lost the last sentence, Andy. Cylinder static. Lost the last sentence. Please read it back."

"All right, Mac. If you see the bomber, *crackle* right away. Got it? If you see *crackle*, call us right away. Got it? Over."

"Got it, Andy. L-4 to Huguenot, over and out."

"Over and out."

The railroad embankment below them went around a wide arc and separated deceptively into two. One of the lines had been pulled up years back, but the marks of the long-ago stacked and burned ties still striped the gravel bed, and it would have been impossible for a stranger to tell from the air whether or not there were any rails running over those marks; terrain from the air can be deceptive unless you know what it is supposed to look like, rather than what it does look like. Martinson, however, knew as well as McDonough which of the two rail spurs was the discontinued one, and banked the Cub in a gentle climbing turn toward the mountain.

The rectangular acres wheeled slowly and solemnly below them, brindled with tiny cows as motionless as toys. After a while the deceptive spur line turned sharply east into a woolly green woods and never came out again. The mountain got larger, the morning ground haze rising up its nearer side, as though the whole forest were smoldering sullenly there.

Martinson turned his head and leaned it back to look out of the corner of one eye at the back seat, but McDonough shook his head. There was no chance at all that the crashed bomber could be on this side of that heavy-shouldered mass of rock.

Martinson shrugged and eased the stick back. The plane bored up into the sky, past four thousand feet, past four thousand, five hundred. Lake Hawthorne passed under the Cub's fat little tires, an irregular sapphire set in the pommel of the mountain. The altimeter crept slowly past five thousand feet; Martinson was taking no chances on being caught in the downdraft on the other side of the hill. At six thousand, he edged the throttle back and leveled out, peering back through the plexiglas.

But there was no sign of any wreck on that side of the mountain, either.

Puzzled, McDonough forced up the top cabin flap on the right side, buttoned it into place against the buffeting slipstream, and thrust his head out into the tearing gale. There was nothing to see on the ground. Straight down, the knife-edge brow of the cliff from which the railroad tracks emerged again drifted slowly away from the Cub's tail; just an inch farther on was the matchbox which was the Otisville siding shack. A sort of shaking of pepper around the matchbox meant people, a small crowd of them—though there was no train due until the Erie's No. 6, which didn't stop at Otisville anyhow.

He thumped Martinson on the shoulder. The adjutant tilted his head back and shouted, "What?"

"Bank right. Something going on around the Otisville station. Go down a bit."

The adjutant jerked out the carburetor-heat toggle and pulled back the throttle. The plane, idling, went into a long, whistling glide along the railroad right of way.

"Can't go too low here," he said. "If we get caught in the downdraft, we'll get slammed right into the mountain."

"I know that. Go on about four miles and make an airline approach back. Then you can climb into the draft. I want to see what's going on down there."

Martinson shrugged and opened the throttle again. The Cub clawed for altitude, then made a half-turn over Howells for the bogus landing run.

The plane went into normal glide and McDonough craned his neck. In a few moments he was able to see what had happened down below. The mountain from this side was steep and sharp; a wounded bomber couldn't possibly have hoped to clear it. At night, on the other hand, the mouth of the railroad tunnel was marked on all three sides, by the lights of the station on the left, the neon sign of the tavern which stood on the brow of the cliff in Otisville (POP. 3,000—HIGH AND HEALTHY) and on the right by the Erie's own signal standard. Radar would have shown the rest: the long regular path of the embankment leading directly into that cul-de-sac of lights, the beetling mass of contours which was the mountain. All these signs would mean "tunnel" in any language.

And the bomber pilot had taken the longest of all possible chances: to come down gliding along the right of way, in the hope of shooting his fuselage cleanly into that tunnel, leaving behind his wings with their dangerous engines and fuel tanks. It was absolutely insane, but that was what he had done.

And, miracle of miracles, he had made it. McDonough could see the wings now, buttered into two-dimensional profiles over the two pilasters of the tunnel. They had hit with such force that the fuel in them must have been vaporized instantly; at least, there was no sign of a fire. And no sign of a fuselage, either.

The bomber's body was inside the mountain, probably halfway or more down the tunnel's one-mile length. It was inconceivable that there could be anything intelligible left of it; but where one miracle has happened, two are possible.

No wonder the little Otisville station was peppered over with the specks of wondering people.

"L-4 to Huguenot. L-4 to Huguenot. Andy, are you there?"

"We read you, Mac. Go ahead."

"We've found your bomber. It's in the Otisville tunnel. Over."

"*Crackle* to L-4. You've lost your mind."

"That's where it is, all the same. We're going to try to make a landing. Send us a team as soon as you can. Out."

"Huguenot to L-4. Don't be a *crackle* idiot, Mac, you can't land there."

"Out," McDonough said. He pounded Martinson's shoulder and gestured urgently downward.

"You want to land?" Martinson said. "Why didn't you say so? We'll never get down on a shallow glide like this." He cleared the engine with a brief burp on the throttle, pulled the Cub up into a sharp stall, and slid off on one wing. The whole world began to spin giddily.

Martinson was losing altitude. McDonough closed his eyes and hung onto his back teeth.

Martinson's drastic piloting got them down to a rough landing, on the wheels, on the road leading to the Otisville station, slightly under a mile away from the mountain. They taxied the rest of the way. The crowd left the mouth of the tunnel to cluster around the airplane the moment it had come to a stop, but a few moments' questioning convinced McDonough that the Otisvilleans knew very little. Some of them had heard "a turrible noise" in the early morning, and with the first light had discovered the bright metal coating the sides of the tunnel. No, there hadn't been any smoke. No, nobody heard any sounds in the tunnel. You couldn't see the other end of it, though; something was blocking it.

"The signal's red on this side," McDonough said thoughtfully while he helped the adjutant tie the plane down. "You used to run the PBX board for the Erie in Port, didn't you, Marty? If you were to phone the station master there, maybe we could get him to throw a block on the other end of the tunnel."

"If there's wreckage in there, the block will be on automatically."

"Sure. But we've got to go in there. I don't want the Number Six piling in after us."

Martinson nodded, and went inside the railroad station. McDonough looked around. There was, as usual, a motorized hand truck parked off the tracks on the other side of the embankment. Many willing hands helped him set it on the right of way, and several huskies got the one-lung engine started for him. Getting his own apparatus out of the plane and onto the truck, however, was a job for which he refused all aid. The stuff was just too delicate, for all its weight, to be allowed in the hands of laymen—and never mind that McDonough himself was almost as much of a layman in neurophysiology as they were; he at least knew the collimating tables and the cookbook.

"O.K.," Martinson said, rejoining them. "Tunnel's blocked at both ends. I talked to Ralph at the dispatcher's; he was steaming—says he's lost four trains already, and another due in from Buffalo in forty-four minutes. We cried a little about it. Do we go now?"

"Right now."

Martinson drew his automatic and squatted down on the front of the truck. The little car growled and crawled toward the tunnel. The spectators murmured and shook their heads knowingly.

Inside the tunnel it was as dark as always, and cold, with a damp chill which struck through McDonough's flight jacket and dungarees. The air was still, and in addition to its musty smell it had a peculiar metallic stench. Thus far, however, there was none of the smell of fuel or of combustion products which McDonough had expected. He found suddenly that he was trembling again, although he did not really believe that the EEG would be needed.

"Did you notice those wings?" Martinson said suddenly, just loud enough to be heard above the popping of the motor. The echoes distorted his voice almost beyond recognition.

"Notice them? What about them?"

"Too short to be bomber wings. Also, no engines."

McDonough swore silently. To have failed to notice a detail as gross as that was a sure sign that he was even more frightened than he had thought. "Anything else?"

"Well, I don't think they were aluminum; too tough. Titanium, maybe, or stainless steel. What have we got in here, anyhow? You *know* the Russkies couldn't get a fighter this far."

There was no arguing that. There was no answering the question, either—not yet.

McDonough unhooked the torch from his belt. Behind them, the white aperture of the tunnel's mouth looked no bigger than a nickel, and the twin bright lines of the rails looked forty miles long. Ahead, the flashlight revealed nothing but the slimy walls of the tunnel, coated with soot.

And then there was a fugitive bluish gleam. McDonough set the motor back down as far as it would go. The truck crawled painfully through the stifling blackness. The thudding of the engine was painful, as though his own heart were trying to move the heavy platform.

The gleam came closer. Nothing moved around it. It was metal, reflecting the light from his torch. Martinson lit his own and brought it into play.

The truck stopped, and there was absolute silence except for the ticking of water on the floor of the tunnel.

"It's a rocket," Martinson whispered. His torch roved over the ridiculously inadequate tail empennage facing them. It was badly crumpled. "In fair shape, considering. At the clip he was going, he must have slammed back and forth like an alarm clapper."

Cautiously they got off the truck and prowled around the gleaming, badly dented spindle. There were clean shears where the wings had been, but the stubs still remained, as though the metal itself had given to the impact before the joints could. That meant welded construction throughout, McDonough remembered vaguely. The vessel rested now roughly in the center of the tunnel, and the railroad tracks had spraddled under its weight. The fuselage bore no identifying marks, except for a red star at the nose; or rather, a red asterisk.

Martinson's torch lingered over the star for a moment, but the adjutant offered no comment. He went around the nose, McDonough trailing.

On the other side of the ship was the death wound; a small, ragged tear in the metal, not far forward of the tail. Some of the raw curls of metal were partially melted. Martinson touched one.

"Flak," he muttered. "Cut his fuel lines. Lucky he didn't blow up."

"How do we get in?" McDonough said nervously. "The cabin didn't even crack. And we can't crawl through that hole."

Martinson thought about it. Then he bent to the lesion in the ship's skin, took a deep breath, and bellowed at the top of his voice:

"*Hey* in there! Open up!"

It took a long time for the echoes to die away. McDonough was paralyzed with pure fright. Anyone of those distorted, ominous rebounding voices could have been an answer. Finally, however, the silence came back.

"So he's dead," Martinson said practically. "I'll bet even his footbones are broken, every one of 'em. Mac, stick your hair net in there and see if you can pick up anything."

"N-not a chance. I can't get anything unless the electrodes are actually t-touching the skull."

"Try it anyhow, and then we can get out of here and let the experts take over. I've about made up my mind it's a missile, anyhow. With this little damage, it could still go off."

McDonough had been repressing that notion since his first sight of the spindle. The attempt to save the fuselage intact, the piloting skill involved, and the obvious cabin windshield all argued against it; but even the bare possibility was somehow twice as terrifying, here under a mountain, as it would have been in the open. With so enormous a mass of rock pressing down on him, and the ravening energies of a sun perhaps waiting to break loose by his side——

No, no; it was a fighter, and the pilot might somehow still be alive. He almost ran to get the electrode net off the truck. He dangled it on its cable inside the flak tear, pulled the goggles over his eyes, and flicked the switch with his thumb.

The Walter goggles made the world inside the tunnel no darker than it actually was, but knowing that he would now be unable to see any gleam of light in the tunnel, should one appear from somewhere—say, in the ultimate glare of hydrogen fusion—increased the pressure of blackness on his brain. Back on the truck the frequency-analyzer began its regular, meaningless peeping, scanning the possible cortical output bands in order of likelihood: First the 0.5 to 3.5 cycles/second band, the delta wave, the last activity of the brain detectable before death; then the four to seven c.p.s. theta channel, the pleasure-scanning waves which went on even during sleep; the alpha rhythm, the visual scanner, at eight to thirteen c.p.s.; the beta rhythms at fourteen to thirty c.p.s. which mirror the tensions of conscious computation, not far below the level of real thought; the gamma band, where——

The goggles lit.

*. . . And still the dazzling sky-blue sheep are grazing in the red field under the rainbow-billed and pea-green birds. . . .*

McDonough snatched the goggles up with a gasp, and stared frantically into the blackness, now swimming with residual images in contrasting colors, melting gradually as the rods and cones in his retina gave up the energy they had absorbed from the scene in the goggles. Curiously, he knew at once where the voice had come from: it had been his mother's reading to him, on Christmas Eve, a story called "A Child's Christmas in Wales." He had not thought of it in well over two decades, but the scene in the toposcope goggles had called it forth irresistibly.

"What's the matter?" Martinson's voice said. "Get anything? Are you sick?"

"No," McDonough muttered. "Nothing."

"Then let's beat it. Do you make a noise like that over nothing every day? My Uncle Crosby did, but then, *he* had asthma."

Tentatively, McDonough lowered the goggles again. The scene came back, still in the same impossible colors, and almost completely without motion. Now that he was able to look at it again, however, he saw that the blue animals were not sheep; they were too large, and they had faces rather like those of kittens. Nor were the enormously slow-moving birds actually birds at all, except that they did seem to be flying—in unlikely straight lines, with slow, mathematically even flappings of unwinglike wings; there was something vegetable about them. The red field was only a dazzling blur, hazing the feet of the blue animals with the huge, innocent kitten's faces. As for the sky, it hardly seemed to be there at all; it was as white as paper.

"Come on," Martinson muttered, his voice edged with irritation. "What's the sense of staying in this hole any more? You bucking for pneumonia?"

"There's . . . something alive in there."

"Not a chance," Martinson said. His voice was noticeably more ragged. "You're dreaming. You said yourself you couldn't pick up——"

"I know what I'm doing," McDonough insisted, watching the scene in the goggles. "There's a live brain in there. Something nobody's ever hit before. It's powerful—no mind in the books ever put out a broadcast like this. It isn't human."

"All the more reason to call in the AF and quit. We can't get in there anyhow. What do you mean, it isn't human? It's a Red, that's all."

"No, it isn't," McDonough said evenly. Now that he thought he knew what they had found, he had stopped trembling. He was still terrified, but it was a different kind of terror: the fright of a man who has at last gotten a clear idea of what it is he is up against. "Human beings just don't broadcast like this. Especially not when they're near dying. And they don't remember huge blue sheep with cat's heads on them, or red grass, or a white sky. Not even if they come from the USSR. Whoever it is in there comes from some place else."

"You read too much. What about the star on the nose?"

McDonough drew a deep breath. "What about it?" he said steadily. "It isn't the insignia of the Red Air Force. I saw that it stopped you, too. No air force I ever heard of flies a red asterisk. It isn't a *cocarde* at all. It's just what it is."

"An asterisk?" Martinson said angrily.

"No, Marty, I think it's a star. A symbol for a *real* star. The AF's gone and knocked us down a spaceship." He pushed the goggles up and carefully withdrew the electrode net from the hole in the battered fuselage.

"And," he said carefully, "the pilot, whatever he is, is still alive—and thinking about home, wherever *that* is."

Though the Air Force had been duly notified by the radio net of McDonough's preposterous discovery, it took its own time about getting a technical crew over to Otisville. It had to, regardless of how much stock it took in the theory. The nearest source of advanced Air Force EEG equipment was just outside Newburgh, at Stewart Field, and it would have to be driven to Otisville by truck; no AF plane slow enough to duplicate Martinson's landing on the road could have handled the necessary payload.

For several hours, therefore, McDonough could do pretty much as he liked with his prize. After only a little urging, Martinson got the Erie dispatcher to send an oxyacetylene torch to the Port Jervis side of the tunnel, on board a Diesel camelback. Persons, who had subsequently arrived in the Aeronca, was all for trying it immediately in the tunnel, but McDonough was restrained by some dim memory of high school experiments with magnesium, a metal which looked very much like this. He persuaded the C.O. to try the torch on the smeared wings first.

The wings didn't burn. They carried the torch into the tunnel, and Persons got to work with it, enlarging the flak hole.

"Is that what-is-it still alive?" Persons asked, cutting steadily.

"I think so," McDonough said, his eyes averted from the tiny sun of the torch. "I've been sticking the electrodes in there about once every five minutes. I get essentially the same picture. But it's getting steadily weaker."

"D'you think we'll reach it before it dies?"

"I don't know. I'm not even sure I want to."

Persons thought that over, lifting the torch from the metal. Then he said, "You've got something there. Maybe I better try that gadget and see what I think."

"No," McDonough said. "It isn't tuned to you."

"Orders, Mac. Let me give it a try. Hand it over."

"It isn't that, Andy. I wouldn't buck you, you know that; you made this squadron. But it's dangerous. Do you want to have an epileptic fit? The chances are nine to five that you would."

"Oh," Persons said. "All right. It's your show." He resumed cutting.

After a while McDonough said, in a remote, emotionless voice: "That's enough. I think I can get through there now, as soon as it cools."

"Suppose there's no passage between the tail and the nose?" Martinson said. "More likely there's a firewall, and we'd never be able to cut through that."

"Probably," McDonough agreed. "We couldn't run the torch near the fuel tanks, anyhow, that's for sure."

"Then what good——"

"If these people think anything like we do, there's bound to be some kind of escape mechanism—something that blows the pilot's capsule free of the ship. I ought to be able to reach it."

"And fire it in *here?*" Persons said. "You'll smash the cabin against the tunnel roof. That'll kill the pilot for sure."

"Not if I disarm it. If I can get the charge out of it, all firing it will do is open the locking devices; then we can take the windshield off and get in. I'll pass the charge out back to you; handle it gently. Let me have your flashlight, Marty, mine's almost dead."

Silently, Martinson handed him the light. He hesitated a moment, listening to the water dripping in the background. Then, with a deep breath, he said, "Well. Here goes nothin'."

He clambered into the narrow opening.

The jungle of pipes, wires and pumps before him was utterly unfamiliar in detail, but familiar in principle. Human beings, given the job of setting up a rocket motor, set it up in this general way. McDonough probed with the light beam, looking for a passage large enough for him to wiggle through.

There didn't seem to be any such passage, but he squirmed his way forward regardless, forcing himself into any opening that presented itself, no matter how small and contorted it seemed. The feeling of entrapment was terrible. If he were to wind up in a cul-de-sac, he would never be able to worm himself backwards out of this jungle of piping . . .

He hit his head a sharp crack on a metal roof, and the metal resounded hollowly. A tank of some kind, empty, or nearly empty. Oxygen? No, unless the stuff had evaporated long ago; the skin of the tank was no colder than any of the other surfaces he had encountered. Propellant, perhaps, or compressed nitrogen—something like that.

Between the tank and what he took to be the inside of the hull, there was a low freeway, just high enough for him to squeeze through if he turned his head sideways. There were occasional supports and ganglions of wiring to be writhed around, but the going was a little better than it had been, back in the engine compartment. Then his head lifted into a slightly larger space, made of walls that curved gently against each other: the front of the tank, he guessed, opposed to the floor of the pilot's capsule and the belly of the hull. Between the capsule and the hull, up rather high, was the outside curve of a tube, large in diameter but very short; it was encrusted with motors, small pumps, and wiring.

An air lock? It certainly looked like one. If so, the trick with the escape mechanism might not have to be worked at all—if indeed the escape device existed.

Finding that he could raise his shoulders enough to rest on his elbows, he studied the wiring. The thickest of the cables emerged from the pilot's capsule; that should be the power line, ready to activate the whole business when the pilot hit the switch. If so, it could be shorted out— provided that there was still any juice in the batteries.

He managed to get the big nippers free of his belt, and dragged forward into a position where he could use them, with considerable straining. He closed their needlelike teeth around the cable and squeezed with all his might. The jaws closed slowly, and the cusps bit in.

There was a deep, surging hum, and all the pumps and motors began to whirr and throb. From back the way he had come, he heard a very muffled distant shout of astonishment.

He hooked the nippers back into his belt and inched forward, raising his back until he was almost curled into a ball. By careful, small movements, as though he were being born, he managed to somersault painfully in the cramped, curved space, and get his head and shoulders back under the tank again, face up this time. He had to trail the flashlight, so that his progress backwards through the utter darkness was as blind as a mole's; but he made it, at long last.

The tunnel, once he had tumbled out into it again, seemed miraculously spacious—almost like flying.

"The damn door opened right up, all by itself," Martinson was chattering. "Scared me green. What'd you do—say 'Open sesame' or something?"

"Yeah," McDonough said. He rescued his electrode net from the hand truck and went forward to the gaping air lock. The door had blocked most of the rest of the tunnel, but it was open wide enough.

It wasn't much of an air lock. As he had seen from inside, it was too short to hold a man; probably it had only been intended to moderate the pressure drop between inside and outside, not prevent such a drop absolutely. Only the outer door had the proper bank-vault heaviness of a true air lock. The inner one, open, was now nothing but a narrow ring of serrated blades, machined to a Johannson-block finish so fine that they were airtight by virtue of molecular cohesion alone—a highly perfected iris diaphragm. McDonough wondered vaguely how the pinpoint hole in the center of the diaphragm was plugged when the iris was fully closed, but his layman's knowledge of engineering failed him entirely there; he could come up with nothing better than a vision of the pilot plugging that hole with a wad of well-chewed bubble gum.

He sniffed the damp, cold, still air. Nothing. If the pilot had breathed anything alien to Earth-normal air, it had already dissipated without trace in the organ pipe of the tunnel. He flashed his light inside the cabin.

The instruments were smashed beyond hope, except for a few at the sides of the capsule. The pilot had smashed them —or rather, his environment had.

Before him in the light of the torch was a heavy, transparent tank of iridescent greenish-brown fluid, with a small figure floating inside it. It had been the tank, which had broken free of its moorings, which had smashed up the rest of the compartment. The pilot was completely enclosed in what looked like an ordinary G-suit, inside the oil; flexible hoses connected to bottles on the ceiling fed him his atmosphere, whatever it was. The hoses hadn't broken, but something inside the G-suit had; a line of tiny bubbles was rising from somewhere near the pilot's neck.

He pressed the EEG electrode net against the tank and looked into the Walter goggles. The sheep with the kitten's faces were still there, somewhat changed in position; but almost all of the color had washed out of the scene. McDonough grunted involuntarily. There was now an atmosphere about the picture which hit him like a blow, a feeling of intense oppression, of intense distress . . .

"Marty," he said hoarsely. "Let's see if we can't cut into that tank from the bottom somehow." He backed down into the tunnel.

"Why? If he's got internal injuries——"

"The suit's been breached. It's filling with that oil from the bottom. If we don't drain the tank, he'll drown first."

"All right. Still think he's a man-from-Mars, Mac?"

"I don't know. It's too small to be a man, you can see that. And the memories aren't like human memories. That's all I know. Can we drill the tank some place?"

"Don't need to," Persons' echo-distorted voice said from inside the air lock. The reflections of his flashlight shifted in the opening like ghosts. "I just found a drain pet cock. Roll up your trouser cuffs, gents."

But the oil didn't drain out of the ship. Evidently it went into storage somewhere inside the hull, to be pumped back into the pilot's cocoon when it was needed again.

It took a long time. The silence came flooding back into the tunnel.

"That oil-suspension trick is neat," Martinson whispered edgily. "Cushions him like a fish. He's got inertia still, but no mass—like a man in free fall."

McDonough fidgeted, but said nothing. He was trying to imagine what the multicolored vision of the pilot could mean. Something about it was nagging at him. It was wrong. Why would a still-conscious and gravely injured pilot be solely preoccupied with remembering the fields of home? Why wasn't he trying to save himself instead—as ingeniously as he had tried to save the ship? He still had electrical power, and in that litter of smashed apparatus which he alone could recognize, there must surely be expedients which still awaited his trial. But he had already given up, though he knew he was dying.

Or did he? The emotional aura suggested a knowledge of things desperately wrong, yet there was no real desperation, no frenzy, hardly any fear—almost as though the pilot did not know what death was, or, knowing it, was confident that it could not happen to him. The immensely powerful, dying mind inside the G-suit seemed curiously uncaring and passive, as though it awaited rescue with supreme confidence —so supreme that it could afford to drift, in an oil-suspended floating dream of home, nostalgic and unhappy, but not really afraid.

And yet it was dying!

"Almost empty," Andy Persons' quiet, garbled voice said into the tunnel.

Clenching his teeth, McDonough hitched himself into the air lock again and tried to tap the fading thoughts on a

higher frequency. But there was simply nothing to hear or
see, though with a brain so strong, there should have been,
at as short a range as this. And it was peculiar, too, that
the visual dream never changed. The flow of thoughts in a
powerful human mind is bewilderingly rapid; it takes weeks
of analysis by specialists before its essential pattern emerges.
This mind, on the other hand, had been holding tenaciously
to this one thought—complicated though it was—for a
minimum of two hours. A truly subidiot performance—being
broadcast with all the drive of a super genius.

Nothing in the cookbook provided McDonough with any
precedent for it.

The suited figure was now slumped against the side of
the empty tank, and the shades inside the toposcope goggles
suddenly began to be distorted with regular, wrenching blurs:
pain waves. A test at the level of the theta waves confirmed
it; the unknown brain was responding to the pain with terrible knots of rage, real blasts of it, so strong and un-
controlled that McDonough could not endure them for more
than a second. His hand was shaking so hard that he could
hardly tune back to the gamma level again.

"We should have left the oil there," he whispered. "We've
moved him too much. The internal injuries are going to kill
him in a few minutes."

"We couldn't let him drown, you said so yourself," Persons
said practically. "Look, there's a seam on this tank that looks
like a torsion seal. If we break it, it ought to open up like
a tired clam. Then we can get him out of here."

As he spoke, the empty tank parted into two shell-like
halves. The pilot lay slumped and twisted at the bottom,
like a doll, his suit glistening in the light of the C.O.'s torch.

"Help me. By the shoulders, real easy. That's it; lift. Easy,
now."

Numbly, McDonough helped. It was true that the oil
would have drowned the fragile, pitiful figure, but this was
no help, either. The thing came up out of the cabin like a
marionette with all its strings cut. Martinson cut the last of
them: the flexible tubes which kept it connected to the ship.
The three of them put it down, sprawling bonelessly.

. . . AND STILL THE DAZZLING SKY-BLUE SHEEP ARE GRAZING
IN THE RED FIELD . . .

Just like that, McDonough saw it.

*A coloring book!*

That was what the scene was. That was why the colors

were wrong, and the size referents. Of course the sheeplike animals did not look much like sheep, which the pilot could never have seen except in pictures. Of course the sheep's heads looked like the heads of kittens; everyone has seen kittens. Of course the brain was powerful out of all proportion to its survival drive and its knowledge of death; it was the brain of a genius, but a genius without experience. And of course, *this* way, the USSR could get a rocket fighter to the United States on a one-way trip.

The helmet fell off the body, and rolled off into the gutter which carried away the water condensing on the wall of the tunnel. Martinson gasped, and then began to swear in a low, grinding monotone. Andy Persons said nothing, but his light, as he played it on the pilot's head, shook with fury.

McDonough, his fantasy of space ships exploded, went back to the hand truck and kicked his tomb-tapping apparatus into small shards and bent pieces. His whole heart was a fuming caldron of pity and grief. He would never knock upon another tomb again.

The blond head on the floor of the tunnel, dreaming its waning dream of a colored paper field, was that of a little girl, barely eight years old.


# King of the Hill

IT DID Col. Hal Gascoigne no good whatsoever to know that he was the only man aboard Satellite Vehicle 1. No good at all. He had stopped reminding himself of the fact some time back.

And now, as he sat sweating in the perfectly balanced air in front of the bombardier board, one of the men spoke to him again:

"Colonel, sir——"

Gascoigne swung around in the seat, and the sergeant—Gascoigne could almost remember the man's name—threw him a snappy Air Force salute.

"Well?"

"Bomb one is primed, sir. Your orders?"

"My orders?" Gascoigne said wonderingly. But the man was already gone. Gascoigne couldn't actually *see* the sergeant leave the control cabin, but he was no longer in it.

While he tried to remember, another voice rang in the

cabin, as flat and razzy as all voices sound on an intercom.

"Radar room. On target."

A regular, meaningless peeping. The timing circuit had cut in.

Or had it? There was nobody in the radar room. There was nobody in the bomb hold, either. There had never been anybody on board SV-1 but Gascoigne, not since he had relieved Grinnell—and Grinnell had flown the station up here in the first place.

Then who had that sergeant been? His name was . . . It was . . .

The hammering of the teletype blanked it out. The noise was as loud as a pom-pom in the echoing metal cave. He got up and coasted across the deck to the machine, gliding in the gravity-free cabin with the ease of a man to whom free fall is almost second nature.

The teletype was silent by the time he reached it, and at first the tape looked blank. He wiped the sweat out of his eyes. There was the message.

MNBVCXZ LKJ HGFDS PYTR AOIU EUIO QPALZM

He got out his copy of *The Well-Tempered Pogo* and checked the speeches of Grundoon the Beaver-Chile for the key letter-sequence on which the code was based. There weren't very many choices. He had the clear in ten minutes.

BOMB ONE WASHINGTON 1700 HRS TAMMANÁNY

There it was. That was what he had been priming the bomb for. But there should have been earlier orders, giving him the go-ahead to prime. He began to rewind the paper.

It was all blank.

And—*Washington?* Why would the Joint Chiefs of Staff order him——

"Colonel Gascoigne, sir."

Gascoigne jerked around and returned the salute. "What's your name?" he snapped.

"Sweeney, sir," the corporal said. Actually it didn't sound very much like Sweeney, or like anything else; it was just a noise. Yet the man's face looked familiar. "Ready with bomb two, sir."

The corporal saluted, turned, took two steps, and faded. He did not vanish, but he did not go out the door, either. He simply receded, became darker and harder to distinguish, and was no longer there. It was as though he and Gascoigne had disagreed about the effects of perspective in the glowing Earthlight, and Gascoigne had turned out to be wrong.

Numbly, he finished rewinding the paper. There was no doubt about it. There the order stood, black on yellow, as plain as plain. Bomb the capital of your own country at 1700 hours. Just incidentally, bomb your own home in the process, but don't give that a second thought. Be thorough, drop two bombs; don't worry about missing by a few seconds of arc and hitting Baltimore instead, or Silver Spring, or Milford, Del. CIG will give you the coordinates, but plaster the area anyhow. That's S.O.P.

With rubbery fingers, Gascoigne began to work the keys of the teletype. Sending on the frequency of Civilian Intelligence Group, he typed:

HELP SHOUT SERIOUS REPEAT SERIOUS PERSONNEL TROUBLE HERE STOP DON'T KNOW HOW LONG I CAN KEEP IT DOWN STOP URGENT GASCOIGNE SV ONE STOP

Behind him, the oscillator peeped rhythmically, timing the drive on the launching rack trunnion.

"Radar room. On target."

Gascoigne did not turn. He sat before the bombardier board and sweated in the perfectly balanced air. Inside his skull, his own voice was shouting:

STOP    STOP    STOP

That, as we reconstructed it afterwards, is how the SV-1 affair began. It was pure luck, I suppose, that Gascoigne sent his message direct to us. Civilian Intelligence Group is rarely called into an emergency when the emergency is just being born. Usually Washington tries to do the bailing job first. Then, when Washington discovers that the boat is still sinking, it passes the bailing can to us—usually with a demand that we transform it into a centrifugal pump, on the double.

We don't mind. Washington's failure to develop a government department similar in function to CIG is the reason why we're in business. The profits, of course, go to Affiliated Enterprises, Inc., the loose corporation of universities and industries which put up the money to build ULTIMAC—and ULTIMAC is, in turn, the reason why Washington comes running to CIG so often.

This time, however, it did not look like the big computer was going to be of much use to us. I said as much to Joan Hadamard, our social sciences division chief, when I handed her the message.

"Um," she said. *"Personnel* trouble? What does he mean? He hasn't got any personnel on that station."

This was no news to me. CIG provided the figures that got the SV-1 into its orbit in the first place, and it was on our advice that it carried only one man. The crew of a space vessel either has to be large or it has to be a lone man; there is no intermediate choice. And SV-1 wasn't big enough to carry a large crew—not to carry them and keep the men from flying at each other's throats sooner or later, that is.

"He means himself," I said. "That's why I don't think this is a job for the computer. It's going to have to be played person-to-person. It's my bet that the man's responsibility-happy; that danger was always implicit in the one-man recommendation."

"The only decent solution is a full complement," Joan agreed. "Once the Pentagon can get enough money from Congress to build a big station."

"What puzzles me is, why did he call us instead of his superiors?"

"That's easy. We process his figures. He trusts us. The Pentagon thinks we're infallible, and he's caught the disease from them."

"That's bad," I said.

"I've never denied it."

"No, what I mean is that it's bad that he called us instead of going through channels. It means that the emergency is at least as bad as he says it is."

I thought about it another precious moment longer while Joan did some quick dialing. As everybody on Earth—with the possible exception of a few Tibetans—already knew, the man who rode SV-1 rode with three hydrogen bombs immediately under his feet—bombs which he could drop with great precision on any spot on the Earth. Gascoigne was, in effect, the sum total of American foreign policy; he might as well have had "Spatial Supremacy" stamped on his forehead.

"What does the Air Force say?" I asked Joan as she hung up.

"They say they're a little worried about Gascoigne. He's a very stable man, but they had to let him run a month over his normal replacement time—why, they don't explain. He's been turning in badly garbled reports over the last week. They're thinking about giving him a dressing down."

"Thinking! They'd better be careful with that stuff, or

they'll hurt themselves. Joan, somebody's going to have to go up there. I'll arrange fast transportation, and tell Gascoigne that help is coming. Who should go?"

"I don't have a recommendation," Joan said. "Better ask the computer."

I did so—on the double.

ULTIMAC said: *Harris.*

"Good luck, Peter," Joan said calmly. Too calmly.

"Yeah," I said. "Or good night."

Exactly what I expected to happen as the ferry rocket approached SV-1, I don't now recall. I had decided that I couldn't carry a squad with me. If Gascoigne was really far gone, he wouldn't allow a group of men to disembark; one man, on the other hand, he might pass. But I suppose I did expect him to put up an argument first.

Nothing happened. He did not challenge the ferry, and he didn't answer hails. Contact with the station was made through the radar automatics, and I was put off on board as routinely as though I was being let into a movie—but a lot more rapidly.

The control room was dark and confusing, and at first I didn't see Gascoigne anywhere. The Earthlight coming through the observation port was brilliant, but beyond the edges of its path the darkness was almost absolute, broken only by the little stars of indicator lenses.

A faint snicking sound turned my eyes in the right direction. There was Gascoigne. He was hunched over the bombardier board, his back to me. In one hand he held a small tool resembling a ticket punch. Its jaws were nibbling steadily at a taut line of tape running between two spools; that had been the sound I'd heard. I recognized the device without any trouble; it was a programmer.

But why hadn't Gascoigne heard me come in? I hadn't tried to sneak up on him, there is no quiet way to come through an air lock anyway. But the punch went on snicking steadily.

"Colonel Gascoigne," I said. There was no answer. I took a step forward. "Colonel Gascoigne, I'm Harris of CIG. What are you doing?"

The additional step did the trick. "Stay away from me," Gascoigne growled, from somewhere way down in his chest. "I'm programming the bomb. Punching in the orders myself. Can't depend on my crew. Stay away."

"Give over for a minute. I want to talk to you."

"That's a new one," said Gascoigne, not moving. "Most of you guys were rushing to set up lauchings before you even reported to me. Who the hell are you, anyhow? There's nobody on board, I know *that* well enough."

"I'm Peter Harris," I said. "From CIG—you called us, remember? You asked us to send help."

"Doesn't prove a thing. Tell me something I *don't* know. Then maybe I'll believe you exist. Otherwise—beat it."

"Nothing doing. Put down that punch."

Gascoigne straightened slowly and turned to look at me. "Well, you don't vanish, I'll give you that," he said. "What did you say your name was?"

"Harris. Here's my ID card."

Gascoigne took the plastic-coated card tentatively, and then removed his glasses and polished them. The gesture itself was perfectly ordinary, and wouldn't have surprised me—except that Gascoigne was not wearing glasses.

"It's hard to see in here," he complained. "Everything gets so steamed up. Hm. All right, you're real. What do you want?"

His finger touched a journal. Silently, the tape began to roll from one spool to another.

"Gascoigne, stop that thing. If you drop any bombs there'll be hell to pay. It's tense enough down below as it is. And there's no reason to bomb anybody."

"Plenty of reason," Gascoigne muttered. He turned toward the teletype, exposing to me for the first time a hip holster cradling a large, black automatic. I didn't doubt that he could draw it with fabulous rapidity, and put the bullets just where he wanted them to go. "I've got orders. There they are. See for yourself."

Cautiously, I sidled over to the teletype and looked. Except for Gascoigne's own message to CIG, and one from Joan Hadamard announcing that I was on my way, the paper was totally blank. There had been no other messages that day unless Gascoigne had changed the roll, and there was no reason why he should have. Those rolls last close to forever.

"When did this order come in?"

"This morning some time. I don't know. Sweeney!" he bawled suddenly, so loud that the paper tore in my hands. "When did that drop order come through?"

Nobody answered. But Gascoigne said almost at once, "There, you heard him."

"I didn't hear anything but you," I said, "and I'm going to stop that tape. Stand aside."

"Not a chance, Mister," Gascoigne said grimly. "The tape rides."

"Who's getting hit?"

"Washington," Gascoigne said, and passed his hand over his face. He appeared to have forgotten the imaginary spectacles.

"That's where your home is, isn't it?"

"It sure is," Gascoigne said. "It sure as hell is, Mister. Cute, isn't it?"

It was cute, all right. The Air Force boys at the Pentagon were going to be given about ten milliseconds to be sorry they'd refused to send a replacement for Gascoigne along with me. *Replace him with who? We can't send his second alternate in anything short of a week. The man has to have retraining, and the first alternate's in the hospital with a ruptured spleen. Besides, Gascoigne's the best man for the job; he's got to be bailed out somehow.*

Sure. With a psychological centrifugal pump, no doubt. In the meantime the tape kept right on running.

"You might as well stop wiping your face, and turn down the humidity instead," I said. "You've already smudged your glasses again."

"Glasses?" Gascoigne muttered. He moved slowly across the cabin, sailing upright like a sea horse, to the blank glass of a closed port. I seriously doubted that he could see his reflection in it, but maybe he didn't really want to see it. "I messed them up, all right. Thanks." He went through the polishing routine again.

A man who thinks he is wearing glasses also thinks he can't see without them. I slid to the programmer and turned off the tape. I was between the spools and Gascoigne now —but I couldn't stay there forever.

"Let's talk a minute, Colonel," I said. "Surely it can't do any harm."

Gascoigne smiled, with a sort of childish craft. "I'll talk," he said. "Just as soon as you start that tape again. I was watching you in the mirror, *before* I took my glasses off."

The liar. I hadn't made a move while he'd been looking into that porthole. His poor pitiful weak old rheumy eyes had seen every move I made while he was polishing his "glasses." I shrugged and stepped away from the programmer.

"You start it," I said. "I won't take the responsibiltiy."

"It's orders," Gascoigne said woodenly. He started the tape running again. "It's their responsibility. What did you want to talk to me about, anyhow?"

"Colonel Gascoigne, have you ever killed anybody?"

He looked startled. "Yes, once I did," he said, almost eagerly. "I crashed a plane into a house. Killed the whole family. Walked away with nothing worse than a burned leg— good as new after a couple of muscle stabilizations. That's what made me shift from piloting to weapons; that leg's not quite good enough to fly with any more."

"Tough."

He snickered suddenly, explosively. "And now look at me," he said. "I'm going to kill my *own* family in a little while. And millions of other people. Maybe the whole world."

How long was "a little while"?

"What have you got against it?" I said.

"Against what—the world? Nothing. Not a damn thing. Look at me; I'm king of the hill up here. I can't complain."

He paused and licked his lips. "It was different when I was a kid," he said. "Not so dull, then. In those days you could get a real newspaper, that you could unfold for the first time yourself, and pick out what you wanted to read. Not like now, when the news comes to you predigested on a piece of paper out of your radio. That's what's the matter with it, if you ask me."

"What's the matter with what?"

"With the news—that's why it's always bad these days. Everything's had something done to it. The milk is homogenized, the bread is sliced, the cars steer themselves, the phonographs will produce sounds no musical instrument could make. Too much meddling, too many people who can't keep their hands off things. Ever fire a kiln?"

"Me?" I said, startled.

"No, I didn't think so. Nobody makes pottery these days. Not by hand. And if they did, who'd buy it? They don't want something that's been made. They want something that's been Done To."

The tape kept on traveling. Down below, there was a heavy rumble, difficult to identify specifically: something heavy being shifted on tracks, or maybe a freight lock opening.

"So now you're going to Do Something to the Earth," I said slowly.

"Not me. It's orders."

"Orders from inside, Colonel Gascoigne. There's nothing on the spools." What else could I do? I didn't have time to take him through two years of psychoanalysis and bring him to his own insight. Besides, I'm not licensed to practice medicine—not on Earth. "I didn't want to say so, but I have to now."

"Say what?" Gascoigne said suspiciously. "That I'm crazy or something?"

"No. I didn't say that. You did," I pointed out. "But I will tell you that that stuff about not liking the world these days is baloney. Or rationalization, if you want a nicer word. You're carrying a screaming load of guilt, Colonel, whether you're aware of it or not."

"I don't know what you're talking about. Why don't you just beat it?"

"No. And you know well enough. You fell all over yourself to tell me about the family you killed in your flying accident." I gave him ten seconds of silence, and then shot the question at him as hard as I could. *What was their name?*

"How do I know? Sweeney or something. Anything. I don't remember."

"Sure you do. Do you think that killing your own family is going to bring the Sweeneys back to life?"

Gascoigne's mouth twisted, but he seemed to be entirely unaware of the grimace. "That's all hogwash," he said. "I never did hold with that psychological claptrap. It's you that's handing out the baloney, not me."

"Then why are you being so vituperative about it? Hogwash, claptrap, baloney—you are working awfully hard to knock it down, for a man who doesn't believe in it."

"Go away," he said sullenly. "I've got my orders. I'm obeying them."

Stalemate. But there was no such thing as stalemate up here. Defeat was the word.

The tape traveled. I did not know what to do. The last bomb problem CIG had tackled had been one we had set up ourselves; we had arranged for a dud to be dropped in New York harbor, to test our own facilities for speed in

determining the nature of the missile. The situation on board
SV-1 was completely different——

Whoa. Was it? Maybe I'd hit something there.

"Colonel Gascoigne," I said slowly, "you might as well
know now that it isn't going to work. Not even if you do get
that bomb off."

"Yes, I can. What's to stop me?" He hooked one thumb
in his belt, just above the holster, so that his fingers tips
rested on the breech of the automatic.

"Your bombs. They aren't alive."

Gascoigne laughed harshly and waved at the controls.
"Tell that to the counter in the bomb hold. Go ahead.
There's a meter you can read, right there on the bombardier
board."

"Sure," I said. "The bombs are radioactive, all right. Have
you ever checked their half life?"

It was a long shot. Gascoigne was a weapons man; if it
were possible to check half life on board the SV-1, he would
have checked it. But I didn't think it was possible.

"What would I do that for?"

"You wouldn't, being a loyal airman. You believe what
your superiors tell you. But I'm a civilian, Colonel. There's
no element in those bombs that will either fuse or fission.
The half life is too long for tritium or for lithium 6, and
it's too short for uranium 235 or radio-thorium. The stuff
is probably strontium 90—in short, nothing but a bluff."

"By the time I finished checking that," Gascoigne said,
"the bomb would be launched anyhow. And you haven't
checked it, either. Try another tack."

"I don't need to. You don't have to believe me. We'll
just sit here and wait for the bomb drop, and then the point
will prove itself. After that, of course, you'll be court-
martialed for firing a wild shot without orders. But since
you're prepared to wipe out your own family, you won't
mind a little thing like twenty years in the guardhouse."

Gascoigne looked at the silently rolling tape. "Sure," he
said, "I've got the orders, anyhow. The same thing would
happen if I didn't obey them. If nobody gets hurt, so much
the better."

A sudden spasm of emotion—I took it to be grief, but
I could have been wrong—shook his whole frame for a
moment. Again, he did not seem to notice it. I said:

"That's right. Not even your family. Of course the whole

world will know the station's a bluff, but if those are the orders——"

"I don't know," Gascoigne said harshly. "I don't know whether I even got any orders. I don't remember where I put them. Maybe they're not real." He looked at me confusedly, and his expression was frighteningly like that of a small boy making a confession.

"You know something?" he said. "I don't know what's real any more. I haven't been able to tell, ever since yesterday. I don't even know if you are real, or your ID card either. What do you think of that?"

"Nothing," I said.

"Nothing! Nothing! That's my trouble. Nothing! I can't tell what's nothing and what's something. You say the bombs are duds. All right. But what if *you're* the dud, and the bombs are real? Answer me that!"

His expression was almost triumphant now.

"The bombs are duds," I said. "And you've gone and steamed up your glasses again. Why don't you turn down the humidity, so you can see for three minutes hand running?"

Gascoigne leaned far forward, so far that he was perilously close to toppling, and peered directly into my face.

"Don't give me that," he said hoarsely. "Don't—give—me that—stuff."

I froze right where I was. Gascoigne watched my eyes for a while. Then, slowly, he put his hand on his forehead and began to wipe it downward. He smeared it over his face, in slow motion, all the way down to his chin.

Then he took the hand away and looked at it, as though it had just strangled him and he couldn't understand why. And finally he spoke.

"It—isn't true," he said dully. "I'm not wearing any glasses. Haven't worn glasses since I was ten. Not since I broke my last pair—playing King of the Hill."

He sat down before the bombardier board and put his head in his hands.

"You win," he said hoarsely. "I must be crazy as a loon. I don't know what I'm seeing and what I'm not. You better take this gun away. If I fired it I might even hit something."

"You're all right," I said. And I meant it; but I didn't waste any time all the same. The automatic first; then the tape. In that order, the sequence couldn't be reversed afterwards.

But the sound of the programmer's journal clicking to "Off" was as loud in that cabin as any gunshot.

"He'll be all right," I told Joan afterwards. "He pulled himself through. I wouldn't have dared to throw it at any other man that fast—but he's got guts."

"Just the same," Joan said, "they'd better start rotating the station captains faster. The next man may not be so tough—and what if *he's* a sleepwalker?"

I didn't say anything. I'd had my share of worries for that week.

"You did a whale of a job yourself, Peter," Joan said. "I just wish we could bank it in the machine. We might need the data later."

"Well, why can't we?"

"The Joint Chiefs of Staff say no. They don't say why. But they don't want any part of it recorded in ULTIMAC —or anywhere else."

I stared at her. At first it didn't seem to make sense. And then it did—and that was worse.

"Wait a minute," I said. "Joan—does that mean what I think it means? Is 'Spatial Supremacy' just as bankrupt as 'Massive Retaliation' was? Is it possible that the satellite— and the bombs . . . Is it possible that I was telling Gascoigne the truth about the bombs being duds?"

Joan shrugged.

"He that darkeneth counsel without wisdom," she said, "isn't earning his salary."


# Common Time

". . . the days went slowly round and round, endless and uneventful as cycles in space. Time, and time-pieces! How many centuries did my hammock tell, as pendulum-like it swung to the ship's dull roll, and ticked the hours and ages."

—Herman Melville, in *Mardi*

*Don't move.*

It was the first thought that came into Garrard's mind when he awoke, and perhaps it saved his life. He lay where he was, strapped against the padding, listening to the round hum of the engines. That in itself was wrong; he should be unable to hear the overdrive at all.

He thought to himself: *Has it begun already?*

Otherwise everything seemed normal. The DFC-3 had crossed over into interstellar velocity, and he was still alive, and the ship was still functioning. The ship should at this moment be traveling at 22.4 times the speed of light—a neat 4,157,000 miles per second.

Somehow Garrard did not doubt that it was. On both previous tries, the ships had whiffed away toward Alpha Centauri at the proper moment when the overdrive should have cut in; and the split second of residual image after they had vanished, subjected to spectroscopy, showed a Doppler shift which tallied with the acceleration predicted for that moment by Haertel.

The trouble was not that Brown and Cellini hadn't gotten away in good order. It was simply that neither of them had ever been heard from again.

Very slowly, he opened his eyes. His eyelids felt terrifically heavy. As far as he could judge from the pressure of the couch against his skin, the gravity was normal; nevertheless, moving his eyelids seemed almost an impossible job.

After long concentration, he got them fully open. The instrument chassis was directly before him, extended over his diaphragm on its elbow joint. Still without moving anything but his eyes—and those only with the utmost patience—he checked each of the meters. Velocity: 22.4 c. Operating temperature: normal. Ship temperature: 37° C. Air pressure: 778 mm. Fuel: No. 1 tank full, No. 2 tank full, No. 3 tank full, No. 4 tank nine tenths full. Gravity: 1 g. Calendar: stopped.

He looked at it closely, though his eyes seemed to focus very slowly, too. It was, of course, something more than a calendar—it was an all-purpose clock, designed to show him the passage of seconds, as well as of the ten months his trip was supposed to take to the double star. But there was no doubt about it: the second hand was motionless.

That was the second abnormality. Garrard felt an impulse to get up and see if he could start the clock again. Perhaps the trouble had been temporary and safely in the past. Immediately there sounded in his head the injunction he had drilled into himself for a full month before the trip had begun—

*Don't move!*

Don't move until you know the situation as far as it can be known without moving. Whatever it was that had snatched Brown and Cellini irretrievably beyond human ken was potent, and totally beyond anticipation. They had both been excellent men, intelligent, resourceful, trained to the point of diminishing returns and not a micron beyond that point—the best men in the Project. Preparations for every knowable kind of trouble had been built into their ships, as they had been built into the DFC-3. Therefore, if there was something wrong nevertheless, it would be something that might strike from some commonplace quarter—and strike only once.

He listened to the humming. It was even and placid, and not very loud, but it disturbed him deeply. The overdrive was supposed to be inaudible, and the tapes from the first unmanned test vehicles had recorded no such hum. The noise did not appear to interfere with the overdrive's operation, or to indicate any failure in it. It was just an irrelevancy for which he could find no reason.

But the reason existed. Garrard did not intend to do so much as draw another breath until he found out what it was.

Incredibly, he realized for the first time that he had not in fact drawn one single breath since he had first come to. Though he felt not the slightest discomfort, the discovery called up so overwhelming a flash of panic that he very nearly sat bolt upright on the couch. Luckily—or so it seemed, after the panic had begun to ebb—the curious lethargy which had affected his eyelids appeared to involve his whole body, for the impulse was gone before he could summon the energy to answer it. And the panic, poignant though it had been for an instant, turned out to be wholly intellectual. In a moment, he was observing that his failure to breathe in no way discommoded him as far as he could tell—it was just there, waiting to be explained . . .

Or to kill him. But it hadn't, yet.

Engines humming; eyelids heavy; breathing absent; calendar stopped. The four facts added up to nothing. The temptation to move something—even if it were only a big toe—was strong, but Garrard fought it back. He had been awake only a short while—half an hour at most—and already had noticed four abnormalities. There were bound to be more, anomalies more subtle than these four; but available to close examination before he had to move. Nor was there anything in particular that he had to do, aside from caring for his own wants; the Project, on the chance that Brown's and Cellini's failure to return had resulted from some tampering with the overdrive, had made everything in the DFC-3 subject only to the computer. In a very real sense, Garrard was just along for the ride. Only when the overdrive was off could he adjust——

*Pock.*

It was a soft, low-pitched noise, rather like a cork coming out of a wine bottle. It seemed to have come just from the right of the control chassis. He halted a sudden jerk of his head on the cushions toward it with a flat fiat of will. Slowly, he moved his eyes in that direction.

He could see nothing that might have caused the sound. The ship's temperature dial showed no change, which ruled out a heat noise from differential contraction or expansion—the only possible explanation he could bring to mind.

He closed his eyes—a process which turned out to be just as difficult as opening them had been—and tried to visualize what the calendar had looked like when he had first come out of anesthesia. After he got a clear and—he was almost sure—accurate picture, Garrard opened his eyes again.

The sound had been the calendar, advancing one second. It was now motionless again, apparently stopped.

He did not know how long it took the second hand to make that jump, normally; the question had never come up. Certainly the jump, when it came at the end of each second, had been too fast for the eye to follow.

Belatedly, he realized what all this cogitation was costing him in terms of essential information. The calendar had moved. Above all and before anything else, he *must* know exactly how long it took it to move again . . .

He began to count, allowing an arbitrary five seconds lost. *One-and-a-six, one-and-a-seven, one-and-an-eight*——

Garrard had gotten only that far when he found himself plunged into hell.

First, and utterly without reason, a sickening fear flooded swiftly through his veins, becoming more and more intense. His bowels began to knot, with infinite slowness. His whole body became a field of small, slow pulses—not so much shaking him as putting his limbs into contrary joggling motions, and making his skin ripple gently under his clothing. Against the hum another sound became audible, a nearly subsonic thunder which seemed to be inside his head. Still the fear mounted, and with it came the pain, and the tenesmus—a boardlike stiffening of his muscles, particularly across his abdomen and his shoulders, but affecting his forearms almost as grievously. He felt himself beginning, very gradually, to double at the middle, a motion about which he could do precisely nothing—a terrifying kind of dynamic paralysis. . . .

It lasted for hours. At the height of it, Garrard's mind, even his very personality, was washed out utterly; he was only a vessel of horror. When some few trickles of reason began to return over that burning desert of reasonless emotion, he found that he was sitting up on the cushions, and that with one arm he had thrust the control chassis back on its elbow so that it no longer jutted over his body. His clothing was wet with perspiration, which stubbornly refused to evaporate or to cool him. And his lungs ached a little, although he could still detect no breathing.

What under God had happened? Was it this that had killed Brown and Cellini? For it would kill Garrard, too—of that he was sure, if it happened often. It would kill him even if it happened only twice more, if the next two such things followed the first one closely. At the very best it would make a slobbering idiot of him; and though the computer might bring Garrard and the ship back to Earth, it would not be able to tell the Project about this tornado of senseless fear.

The calendar said that the eternity in hell had taken three seconds. As he looked at it in academic indignation, it said *pock* and condescended to make the total seizure four seconds long. With grim determination, Garrard began to count again.

He took care to establish the counting as an absolutely even, automatic process which would not stop at the back of his mind no matter what other problem he tackled along with it, or what emotional typhoons should interrupt him. Really compulsive counting cannot be stopped by anything— not the transports of love nor the agonies of empires. Garrard knew the dangers in deliberately setting up such a mechanism in his mind, but he also knew how desperately he needed to time that clock tick. He was beginning to understand what had happened to him—but he needed exact measurement before he could put that understanding to use.

Of course there had been plenty of speculation on the possible effect of the overdrive on the subjective time of the pilot, but none of it had come to much. At any speed below the velocity of light, subjective and objective time were

exactly the same as far as the pilot was concerned. For an
observer on Earth, time aboard the ship would appear to be
vastly slowed at near-light speeds; but for the pilot himself
there would be no apparent change.

Since flight beyond the speed of light was impossible—
although for slightly differing reasons—by both the current
theories of relativity, neither theory had offered any clue as
to what would happen on board a translight ship. They would
not allow that any such ship could even exist. The Haertel
transformation, on which, in effect, the DFC-3 flew, was
nonrelativistic: it showed that the apparent elapsed time of a
translight journey should be identical in ship-time, and in the
time of observers at both ends of the trip.

But since ship and pilot were part of the same system,
both covered by the same expression in Haertel's equation,
it had never occurred to anyone that the pilot and the ship
might keep different times. The notion was ridiculous.

*One-and-a-sevenhundredone, one-and-a-sevenhundredtwo,
one - and - a - sevenhundredthree, one - and - a - sevenhundred
four . . .*

The ship was keeping ship-time, which was identical with
observer-time. It would arrive at the Alpha Centauri system
in ten months. But the pilot was keeping Garrard-time, and
it was beginning to look as though he wasn't going to arrive
at all.

It was impossible, but there it was. Something—almost
certainly an unsuspected physiological side effect of the overdrive field on human metabolism, an effect which naturally
could not have been detected in the preliminary, robot-
piloted tests of the overdrive—had speeded up Garrard's
subjective apprehension of time, and had done a thorough
job of it.

The second hand began a slow, preliminary quivering as
the calendar's innards began to apply power to it. *Seventy-
hundred-forty-one, seventy-hundred-forty-two, seventy-hun-
dred-forty-three . . .*

At the count of 7,058 the second hand began the jump to
the next graduation. It took it several apparent minutes to get
across the tiny distance, and several more to come completely to rest. Later still, the sound came to him:

*pock.*

In a fever of thought, but without any real physical agitation, his mind began to manipulate the figures. Since it took
him longer to count an individual number as the number be-

came larger, the interval between the two calendar ticks probably was closer to 7,200 seconds than to 7,058. Figuring backward brought him quickly to the equivalence he wanted:

One second in ship-time was two hours in Garrard-time.

Had he really been counting for what was, for him, two whole hours? There seemed to be no doubt about it. It looked like a long trip ahead.

Just how long it was gong to be struck him with stunning force. Time had been slowed for him by a factor of 7200. He would get to Alpha Centauri in just 72,000 months.

Which was—

*Six thousand years!*

## 2

Garrard sat motionless for a long time after that, the Nessus-shirt of warm sweat swathing him persistently, refusing even to cool. There was, after all, no hurry.

Six thousand years. There would be food and water and air for all that time, or for sixty or six hundred thousand years; the ship would synthesize his needs, as a matter of course, for as long as the fuel lasted, and the fuel bred itself. Even if Garrard ate a meal every three seconds of objective, or ship, time (which, he realized suddenly, he wouldn't be able to do, for it took the ship several seconds of objective time to prepare and serve up a meal once it was ordered; he'd be lucky if he ate once a day, Garrard-time), there would be no reason to fear any shortage of supplies. That had been one of the earliest of the possibilities for disaster that the Project engineers had ruled out in the design of the DFC-3.

But nobody had thought to provide a mechanism which would indefinitely refurbish Garrard. After six thousand years, there would be nothing left of him but a faint film of dust on the DFC-3's dully gleaming horizontal surfaces. His corpse might outlast him a while, since the ship itself was sterile—but eventually he would be consumed by the bacteria which he carried in his own digestive tract. He needed those bacteria to synthesize part of his B-vitamin needs while he lived, but they would consume him without compunction once he had ceased to be as complicated and delicately balanced a thing as a pilot—or as any other kind of life.

Garrard was, in short, to die before the DFC-3 had gotten fairly away from Sol; and when, after 12,000 apparent years, the DFC-3 returned to Earth, not even his mummy would be still aboard.

The chill that went through him at that seemed almost unrelated to the way he thought he felt about the discovery; it lasted an enormously long time, and insofar as he could characterize it at all, it seemed to be a chill of urgency and excitement—not at all the kind of chill he should be feeling at a virtual death sentence. Luckily it was not as intolerably violent as the last such emotional convulsion; and when it was over, two clock ticks later, it left behind a residuum of doubt.

Suppose that this effect of time-stretching was only mental? The rest of his bodily processes might still be keeping ship-time; Garrard had no immediate reason to believe otherwise. If so, he would be able to move about only on ship-time, too; it would take many apparent months to complete the simplest task.

But he would live, if that were the case. His mind would arrive at Alpha Centauri six thousand years older, and perhaps madder, than his body, but he would live.

If, on the other hand, his bodily movements were going to be as fast as his mental processes, he would have to be enormously careful. He would have to move slowly and exert as little force as possible. The normal human hand movement, in such a task as lifting a pencil, took the pencil from a state of rest to another state of rest by imparting to it an acceleration of about two feet per second per second—and, of course, decelerated it by the same amount. If Garrard were to attempt to impart to a two-pound weight, which was keeping ship-time, an acceleration of 14,440 ft/sec² in his time, he'd have to exert a force of 900 pounds on it.

The point was not that it couldn't be done—but that it would take as much effort as pushing a stalled jeep. He'd never be able to lift that pencil with his forearm muscles alone; he'd have to put his back into the task.

And the human body wasn't engineered to maintain stresses of that magnitude indefinitely. Not even the most powerful professional weight-lifter is forced to show his prowess throughout every minute of every day.

*Pock.*

That was the calendar again; another second had gone by. Or another two hours. It had certainly seemed longer than a second, but less than two hours, too. Evidently subjective time was an intensively recomplicated measure. Even in this world of micro-time—in which Garrard's mind, at least, seemed to be operating—he could make the lapses between calendar ticks seem a little shorter by becoming actively interested in some problem or other. That would help, during the waking hours, but it would help only if the rest of his body were *not* keeping the same time as his mind. If it were not, then he would lead an incredibly active, but perhaps not intolerable, mental life during the many centuries of his awake-time, and would be mercifully asleep for nearly as long.

Both problems—that of how much force he could exert with his body, and how long he could hope to be asleep in his mind—emerged simultaneously into the forefront of his consciousness while he still sat inertly on the hammock, their terms still much muddled together. After the single tick of the calendar, the ship—or the part of it that Garrard could see from here—settled back into complete rigidity. The sound of the engines, too, did not seem to vary in frequency or amplitude, at least as far as his ears could tell. He was still not breathing. Nothing moved, nothing changed.

It was the fact that he could still detect no motion of his diaphragm or his rib cage that decided him at last. His body had to be keeping ship-time, otherwise he would have blacked out from oxygen starvation long before now. That assumption explained, too, those two incredibly prolonged, seemingly sourceless saturnalias of emotion through which he had suffered: they had been nothing more nor less than the response of his endocrine glands to the purely intellectual reactions he had experienced earlier. He had discovered that he was not breathing, had felt a flash of panic and had tried to sit up. Long after his mind had forgotten those two impulses, they had inched their way from his brain down his nerves to the glands and muscles involved, and actual, *physical* panic had supervened. When that was over, he actually *was* sitting up, though the flood of adrenalin had prevented his noticing the motion as he had made it. The later chill—less violent, and apparently associated with the discovery that he might die long before the trip was completed—actually had been his body's response to a much earlier mental command—the abstract fever of interest he had felt while computing the time differential had been responsible for it.

Obviously, he was going to have to be very careful with apparently cold and intellectual impulses of any kind—or he would pay for them later with a prolonged and agonizing glandular reaction. Nevertheless, the discovery gave him considerable satisfaction, and Garrard allowed it free play; it certainly could not hurt him to feel pleased for a few hours, and the glandular pleasure might even prove helpful if it caught him at a moment of mental depression. Six thousand years, after all, provided a considerable number of opportunities for feeling down in the mouth; so it would be best to encourage all pleasure moments, and let the after-reaction last as long as it might. It would be the instants of panic, of fear, of gloom, which he would have to regulate sternly the moment they came into his mind; it would be those which would otherwise plunge him into four, five, six, perhaps even ten, Garrard-hours of emotional inferno.

*Pock.*

There now, that was very good: there had been two Garrard-hours which he had passed with virtually no difficulty of any kind, and without being especially conscious of their passage. If he could really settle down and become used to this kind of scheduling, the trip might not be as bad as he had at first feared. Sleep would take immense bites out of it; and during the waking periods he could put in one hell of a lot of creative thinking. During a single day of ship time, Garrard could get in more thinking than any philosopher of Earth could have managed during an entire lifetime. Garrard could, if he disciplined himself sufficiently, devote his mind for a century to running down the consequences of a single thought, down to the last detail, and still have millennia left to go on to the next thought. What panoplies of pure reason could he not have assembled by the time 6,000 years had gone by? With sufficient concentration, he might come up with the solution to the Problem of Evil between breakfast and dinner of a single ship's day, and in a ship's month might put his finger on the First Cause!

*Pock.*

Not that Garrard was sanguine enough to expect that he would remain logical or even sane throughout the trip. The vista was still grim, in much of its detail. But the opportunities, too, were there. He felt a momentary regret that it hadn't been Haertel, rather than himself, who had been given such an opportunity—

*Pock.*

—for the old man could certainly have made better use of it than Garrard could. The situation demanded someone trained in the highest rigors of mathematics to be put to the best conceivable use. Still and all Garrard began to feel—

*Pock.*

—that he would give a good account of himself, and it tickled him to realize that (as long as he held onto his essential sanity) he would return—

*Pock.*

—to Earth after ten Earth months with knowledge centuries advanced beyond anything—

*Pock.*

—that Haertel knew, or that anyone could know—

*Pock.*

—who had to work within a normal lifetime. *Pck.* The whole prospect tickled him. *Pck.* Even the clock tick seemed more cheerful. *Pck.* He felt fairly safe now *Pck* in disregarding his drilled-in command *Pck* against moving *Pck*, since in any *Pck* event he *Pck* had already *Pck* moved *Pck* without *Pck* being *Pck* harmed *Pck* Pck Pck Pck Pck *pckpckpckpck-pckpckpck.* . . .

He yawned, stretched, and got up. It wouldn't do to be too pleased, after all. There were certainly many problems that still needed coping with, such as how to keep the impulse toward getting a ship-time task performed going, while his higher centers were following the ramifications of some purely philosophical point. And besides . . .

*And besides, he had just moved.*

More than that; he had just performed a complicated maneuver with his body *in normal time!*

Before Garrard looked at the calendar itself, the message it had been ticking away at him had penetrated. While he had been enjoying the protracted, glandular backwash of his earlier feeling of satisfaction, he had failed to notice, at least consciously, that the calendar was accelerating.

Good-bye, vast ethical systems which would dwarf the Greeks. Good-bye, calculuses aeons advanced beyond the spinor calculus of Dirac. Good-bye, cosmologies by Garrard which would allot the Almighty a job as third-assistant-waterboy in an $n$-dimensional backfield.

Good-bye, also, to a project he had once tried to undertake in college—to describe and count the positions of love, of which, according to under-the-counter myth, there were supposed to be at least forty eight. Garrard had never been able to carry his tally beyond twenty, and he had just lost what was probably his last opportunity to try again.

The micro-time in which he had been living had worn off, only a few objective minutes after the ship had gone into overdrive and he had come out of the anesthetic. The long intellectual agony, with its glandular counterpoint, had come to nothing. Garrard was now keeping ship-time.

Garrard sat back down on the hammock, uncertain whether to be bitter or relieved. Neither emotion satisfied him in the end; he simply felt unsatisfied. Micro-time had been bad enough while it lasted; but now it was gone, and everything seemed normal. How could so transient a thing have killed Brown and Cellini? They were stable men, more stable, by his own private estimation, than Garrard himself. Yet he had come through it. Was there more to it than this?

And if there was—what, conceivably, could it be?

There was no answer. At his elbow, on the control chassis which he had thrust aside during that first moment of infinitely protracted panic, the calendar continued to tick. The engine noise was gone. His breath came and went in natural rhythm. He felt light and strong. The ship was quiet, calm, unchanging.

The calendar ticked, faster and faster. It reached and passed the first hour, ship-time, of flight in overdrive.

*Pock.*

Garrard looked up in surprise. The familiar noise, this time, had been the hour-hand jumping one unit. The minute-hand was already sweeping past the past half-hour. The second-hand was whirling like a propellor—and while he watched it, it speeded up to complete invisibility—

*Pock.*

Another hour. The half-hour already passed. *Pock.* Another hour. *Pock.* Another. *Pock. Pock. Pock, Pock, Pock, Pock, pck-pck-pck-pck-pckpckpckpck.* . . .

The hands of the calendar swirled toward invisibility as time ran away with Garrard. Yet the ship did not change. It stayed there, rigid, inviolate, invulnerable. When the date tumblers reached a speed at which Garrard could no longer read them, he discovered that once more he could not move —and that, although his whole body seemed to be aflutter like that of a hummingbird, nothing coherent was coming to him through his senses. The room was dimming, becoming redder; or no, it was . . .

But he never saw the end of the process, never was

allowed to look from the pinnacle of macro-time toward which the Haertel overdrive was taking him.

Pseudo-death took him first.

## 3

That Garrard did not die completely, and within a comparatively short time after the DFC-3 had gone into overdrive, was due to the purest of accidents; but Garrard did not know that. In fact, he knew nothing at all for an indefinite period, sitting rigid and staring, his metabolism slowed down to next to nothing, his mind almost utterly inactive. From time to time, a single wave of low-level metabolic activity passed through him—what an electrician might have termed a "maintenance turnover"—in response to the urgings of some occult survival urge; but these were of so basic a nature as to reach his consciousness not at all. This was the pseudo-death.

When the observer actually arrived, however, Garrard woke. He could make very little sense out of what he saw or felt even now; but one fact was clear: the overdrive was off—and with it the crazy alterations in time rates—and there was strong light coming through one of the ports. The first leg of the trip was over. It had been these two changes in his environment which had restored him to life.

The thing (or things) which had restored him to consciousness, however, was—it was what? It made no sense. It was a construction, a rather fragile one, which completely surrounded his hammock. No, it wasn't a construction, but evidently something alive—a living being, organized horizontally, that had arranged itself in a circle about him. No, it was a number of beings. Or a combination of all of these things.

How it had gotten into the ship was a mystery, but there it was. Or there they were.

"How do you hear?" the creature said abruptly. Its voice, or their voices, came at equal volume from every point in the circle, but not from any particular point in it. Garrard could think of no reason why that should be unusual.

"I—" he said. "Or we—we hear with our ears. Here."

His answer, with its unintentionally long chain of open vowel sounds, rang ridiculously. He wondered why he was speaking such an odd language.

"We-they wooed to pitch you-yours thiswise," the creature said. With a thump, a book from the DFC-3's ample library fell to the deck beside the hammock. "We wooed there and there and there for a many. You are the being-Garrard. We-they are the clinesterton beademung, with all of love."

"With all of love," Garrard echoed. The beademung's use of the language they both were speaking was odd; but again Garrard could find no logical reason why the beademung's usage should be considered wrong.

"Are—are you-they from Alpha Centauri?" he said hesitantly.

"Yes, we hear the twin radioceles, that show there beyond the gift-orifices. We-they pitched that the being-Garrard with most adoration these twins and had mind to them, soft and loud alike. How do you hear?"

This time the being-Garrard understood the question. "I hear Earth," he said. "But that is very soft, and does not show."

"Yes," said the beademung. "It is a harmony, not a first, as ours. The All-Devouring listens to lovers there, not on the radioceles. Let me-mine pitch you-yours so to have mind of the rodalent beademung and other brothers and lovers, along the channel which is fragrant to the being-Garrard."

Garrard found that he understood the speech without difficulty. The thought occurred to him that to understand a language on its own terms—without having to put it back into English in one's own mind—is an ability that is won only with difficulty and long practice. Yet, instantly his mind said, "But it *is* English," which of course it was. The offer the clinesterton beademung had just made was enormously hearted, and he in turn was much minded and of love, to his own delighting as well as to the beademungen; that almost went without saying.

There were many matings of ships after that, and the being-Garrard pitched the harmonies of the beademungen, leaving his ship with the many gift orifices in harmonic for the All-Devouring to love, while the beademungen made show of they-theirs.

He tried, also, to tell how he was out of love with the overdrive, which wooed only spaces and times, and made featurelings. The rodalent beademung wooed the overdrive, but it did not pitch he-them.

Then the being-Garrard knew that all the time was devoured, and he must hear Earth again.

"I pitch you-them to fullest love," he told the beademungen, "I shall adore the radioceles of Alpha and Proxima Centauri, 'on Earth as it is in Heaven.' Now the overdrive my-other must woo and win me, and make me adore a featureling much like silence."

"But you will be pitched again," the clinesterton beademung said. "After you have adored Earth. You are much loved by Time, the All-Devouring. We-they shall wait for this othering."

Privately Garrard did not faith as much, but he said, "Yes, we-they will make a new wooing of the beademungen at some other radiant. With all of love."

On this the beademungen made and pitched adorations, and in the midst the overdrive cut in. The ship with the many gift orifices and the being-Garrard him-other saw the twin radioceles sundered away.

Then, once more, came the pseudo-death.

4

When the small candle lit in the endless cavern of Garrard's pseudo-dead mind, the DFC-3 was well inside the orbit of Uranus. Since the sun was still very small and distant, it made no spectacular display through the nearby port, and nothing called him from the post-death sleep for nearly two days.

The computers waited patiently for him. They were no longer immune to his control; he could now tool the ship back to Earth himself if he so desired. But the computers were also designed to take into account the fact that he might be truly dead by the time the DFC-3 got back. After giving him a solid week, during which time he did nothing but sleep, they took over again. Radio signals began to go out, tuned to a special channel.

An hour later, a very weak signal came back. It was only a directional signal, and it made no sound inside the DFC-3 —but it was sufficient to put the big ship in motion again.

It was that which woke Garrard. His conscious mind was still glazed over with the icy spume of the pseudo-death; and as far as he could see the interior of the cabin had not changed one whit, except for the book on the deck—

The book. The clinesterton beademung had dropped it there. But what under God was a clinesterton beademung? And what was he, Garrard, crying about? It didn't make sense. He remembered dimly some kind of experience out there by the Centauri twins—

*—the twin radioceles—*

There was another one of those words. It seemed to have Greek roots, but he knew no Greek—and besides, why would Centaurians speak Greek?

He leaned forward and actuated the switch which would roll the shutter off the front port, actually a telescope with a translucent viewing screen. It showed a few stars, and a faint nimbus off on one edge which might be the Sun. At about one o'clock on the screen, was a planet about the size of a pea which had tiny projections, like teacup handles, on each side. The DFC-3 hadn't passed Saturn on its way out; at that time it had been on the other side of the Sun from the route the starship had had to follow. But the planet was certainly difficult to mistake.

Garrard was on his way home—and he was still alive and sane. Or was he still sane? These fantasies about Centaurians—which still seemed to have such a profound emotional effect upon him—did not argue very well for the stability of his mind.

But they were fading rapidly. When he discovered, clutching at the handiest fragments of the "memories," that the plural of *beademung* was *beademungen*, he stopped taking the problem seriously. Obviously a race of Centaurians who spoke Greek wouldn't also be forming weak German plurals. The whole business had obviously been thrown up by his unconscious.

But what *had* he found by the Centaurus stars?

There was no answer to that question but that incomprehensible garble about love, the All-Devouring, and beademungen. Possibly, he had never seen the Centaurus stars at all, but had been lying here, cold as a mackerel, for the entire twenty months.

*Or had it been 12,000 years?* After the tricks the overdrive had played with time, there was no way to tell what the objective date actually was. Frantically Garrard put the telescope into action. Where was the Earth? After 12,000 years—

The Earth was there. Which, he realized swiftly, proved nothing. The Earth had lasted for many millions of years; 12,000 years was nothing to a planet. The Moon was there, too; both were plainly visible, on the far side of the Sun—but not too far to pick them out clearly, with the telescope at highest power. Garrard could even see a clear sun-highlight on the Atlantic Ocean, not far east of Greenland; evidently the computers were bringing the DFC-3 in on the Earth from about 23° north of the plane of the ecliptic.

The Moon, too, had not changed. He could even see on its face the huge splash of white, mimicking the sun-highlight on Earth's ocean, which was the magnesium hydroxide landing beacon, which had been dusted over the Mare Vaporum in the earliest days of space flight, with a dark spot on its southern edge which could only be the crater Monilius.

But that again proved nothing. The Moon never changed. A film of dust laid down by modern man on its face would last for millennia—what, after all, existed on the Moon to blow it away? The Mare Vaporum beacon covered more than 4,000 square miles; age would not dim it, nor could man himself undo it—either accidentally, or on purpose—in anything under a century. When you dust an area that large on a world without atmosphere, it stays dusted.

He checked the stars against his charts. They hadn't moved; why should they have, in only 12,000 years? The pointer stars in the Dipper still pointed to Polaris. Draco, like a fantastic bit of tape, wound between the two Bears, and Cepheus and Cassiopeia, as it always had done. These constellations told him only that it was spring in the northern hemisphere of Earth.

*But spring of what year?*

Then, suddenly, it occurred to Garrard that he had a method of finding the answer. The Moon causes tides in the Earth, and action and reaction are always equal and opposite. The Moon cannot move things on Earth without itself being affected—and that effect shows up in the moon's angular momentum. The Moon's distance from the Earth increases steadily by 0.6 inches every year. At the end of 12,000 years, it should be 600 feet farther away from the Earth.

Was it possible to measure? Garrard doubted it, but he got out his ephemeris and his dividers anyhow, and took pictures. While he worked, the Earth grew nearer. By the time he had finished his first calculation—which was indecisive, because it allowed a margin for error greater than the distances he was trying to check—Earth and Moon were close enough in the telescope to permit much more accurate measurements.

Which were, he realized wryly, quite unnecessary. The computer had brought the DFC-3 back, not to an observed sun or planet, but simply to a calculated point. That Earth and Moon would not be near that point when the DFC-3 returned was not an assumption that the computer could make. That the Earth was visible from here was already good and sufficient proof that no more time had elapsed than had been calculated for from the beginning.

This was hardly new to Garrard; it had simply been retired to the back of his mind. Actually he had been doing all this figuring for one reason, and one reason only: because deep in his brain, set to work by himself, there was a mechanism that demanded counting. Long ago, while he was still trying to time the ship's calendar, he had initiated compulsive counting—and it appeared that he had been counting ever since. That had been one of the known dangers of deliberately starting such a mental mechanism; and now it was bearing fruit in these perfectly useless astronomical exercises.

The insight was healing. He finished the figures roughly, and that unheard moron deep inside his brain stopped counting at last. It had been pawing its abacus for twenty months now, and Garrard imagined that it was as glad to be retired as he was to feel it go.

His radio squawked, and said anxiously, "DFC-3, DFC-3. Garrard, do you hear me? Are you still alive? Everybody's going wild down here. Garrard, if you hear me, call us!"

It was Haertel's voice. Garrard closed the dividers so convulsively that one of the points nipped into the heel of his hand. "Haertel, I'm here. DFC-3 to the Project. This is Garrard." And then, without knowing quite why, he added: "With all of love."

Haertel, after all the hoopla was over, was more than interested in the time effects. "It certainly enlarges the manifold in which I was working," he said. "But I think we can account for it in the transformation. Perhaps even factor it out, which would eliminate it as far as the pilot is concerned. We'll see, anyhow."

Garrard swirled his highball reflectively. In Haertel's cramped old office, in the Project's administration shack, he felt both strange and as old, as compressed, constricted. He said, "I don't think I'd do that, Adolph. I think it saved my life."

"How?"

"I told you that I seemed to die after a while. Since I got

home, I've been reading; and I've discovered that the psychologists take far less stock in the individuality of the human psyche than you and I do. You and I are physical scientists, so we think about the world as being all outside our skins—something which is to be observed, but which doesn't alter the essential *I*. But evidently, that old solipsistic position isn't quite true. Our very personalities, really, depend in large part upon *all* the things in our environment, large and small, that exist outside our skins. If by some means you could cut a human being off from every sense impression that comes to him from outside, he would cease to exist as a personality within two or three minutes. Probably he would die."

"Unquote: Harry Stack Sullivan," Haertel said, dryly. "So?"

"So," Garrard said, "think of what a monotonous environment the inside of a spaceship is. It's perfectly rigid, still, unchanging, lifeless. In ordinary interplanetary flight, in such an environment, even the most hardened spaceman may go off his rocker now and then. You know the typical spaceman's psychosis as well as I do, I suppose. The man's personality goes rigid, just like his surroundings. Usually he recovers as soon as he makes port, and makes contact with a more-or-less normal world again.

"But in the DFC-3, I was cut off from the world around me much more severely. I couldn't look outside the ports—I was in overdrive, and there was nothing to see. I couldn't communicate with home, because I was going faster than light. And then I found I couldn't move either, for an enormous long while; and that even the instruments that are in constant change for the usual spaceman wouldn't be in motion for me. Even those were fixed.

"After the time rate began to pick up, I found myself in an even more impossible box. The instruments moved, all right, but then they moved too *fast* for me to read them. The whole situation was now utterly rigid—and, in effect, I died. I froze as solid as the ship around me, and stayed that way as long as the overdrive was on."

"By that showing," Haertel said dryly, "the time effects were hardly your friends."

"But they were, Adolph. Look. Your engines act on subjective time; they keep it varying along continuous curves—from far-too-slow to far-too-fast—and, I suppose, back down again. Now, this is a *situation of continuous change*. It wasn't marked enough, in the long run, to keep me out of pseudo-death; but it was sufficient to protect me from being obliterated altogether, which I think is what happened to Brown and Cellini. Those men knew that they could shut down the overdrive if they could just get to it, and they killed themselves trying. But I knew that I just had to sit and take it—and, by my great good luck, your sine-curve time variation made it possible for me to survive."

"Ah, ah," Haertel said. "A point worth considering—though I doubt that it will make interstellar travel very popular!"

He dropped back into silence, his thin mouth pursed. Garrard took a grateful pull at his drink.

At last Haertel said: "Why are you in trouble over these Centaurians? It seems to me that you have done a good job. It was nothing that you were a hero—any fool can be brave—but I see also that you *thought,* where Brown and Cellini evidently only reacted. Is there some secret about what you found when you reached those two stars?"

Garrard said, "Yes, there is. But I've already told you what it is. When I came out of the pseudo-death, I was just a sort of plastic palimpsest upon which anybody could have made a mark. My own environment, my ordinary Earth environment, was a hell of a long way off. My present surroundings were nearly as rigid as they had ever been. When I met the Centaurians—if I did, and I'm not at all sure of that—*they* became the most important thing in my world, and my personality changed to accommodate and understand them. That was a change about which I couldn't do a thing.

"Possibly I did understand them. But the man who understood them wasn't the same man you're talking to now, Adolph. Now that I'm back on Earth, I don't understand that man. He even spoke English in a way that's gibberish to me. If I can't understand myself during that period—and I can't; I don't even believe that that man was the Garrard I know—what hope have I of telling you or the Project about the Centurians? They found me in a controlled environment, and they altered me by entering it. Now that they're gone, nothing comes through; I don't even understand why I think they spoke English!"

"Did they have a name for themselves?"

"Sure," Garrard said. "They were the beademungen."

"What did they look like?"

"I never saw them."

Haertel leaned forward. "Then . . ."

"I heard them. I think." Garrard shrugged, and tasted his Scotch again. He was home, and on the whole he was pleased.

But in his malleable mind he heard someone say, *On Earth, as it is in Heaven;* and then, in another voice, which might also have been his own (why had he thought "him-other"?), *It is later than you think.*

"Adolph," he said, "is this all there is to it? Or are we going to go on with it from here? How long will it take to make a better starship, a DFC-4?"

"Many years," Haertel said, smiling kindly. "Don't be anxious, Garrard. You've come back, which is more than the others managed to do, and nobody will ask you to go out again. I really think that it's hardly likely that we'll get another ship built during your lifetime; and even if we do, we'll be slow to launch it. We really have very little information about what kind of playground you found out there."

"I'll go," Garrard said. "I'm not afraid to go back—I'd like to go. Now that I know how the DFC-3 behaves, I could take it out again, bring you back proper maps, tapes, photos."

"Do you really think," Haertel said, his face suddenly serious, "that we could let the DFC-3 go out again? Garrard, we're going to take that ship apart practically molecule by molecule; that's preliminary to the building of any DFC-4. And no more can we let you go. I don't mean to be cruel, but has it occurred to you that this desire to go back may be the result of some kind of post-hypnotic suggestion? If so, the more badly you want to go back, the more dangerous to us all you may be. We are going to have to examine you just as thoroughly as we do the ship. If these beademungen wanted you to come back, they must have had a reason— and we have to know that reason."

Garrard nodded, but he knew that Haertel could see the slight movement of his eyebrows and the wrinkles forming in his forehead, the contractions of the small muscles which stop the flow of tears only to make grief patent on the rest of the face.

"In short," he said, *"don't move."*

Haertel looked politely puzzled. Garrard, however, could

say nothing more. He had returned to humanity's common time, and would never leave it again.

Not even, for all his dimly remembered promise, with all there was left in him of love.

# A Work of Art

INSTANTLY, he remembered dying. He remembered it, however, as if at two removes—as though he were remembering a memory, rather than an actual event; as though he himself had not really been there when he died.

Yet the memory was all from his own point of view, not that of some detached and disembodied observer which might have been his soul. He had been most conscious of the rasping, unevenly drawn movements of the air in his chest. Blurring rapidly, the doctor's face had bent over him, loomed, come closer, and then had vanished as the doctor's head passed below his cone of vision, turned sideways to listen to his lungs.

It had become rapidly darker, and then, only then, had he realized that these were to be his last minutes. He had tried dutifully to say Pauline's name, but his memory contained no record of the sound—only of the rattling breath, and of the film of sootiness thickening in the air, blotting out everything for an instant.

Only an instant, and then the memory was over. The room was bright again, and the ceiling, he noticed with wonder, had turned a soft green. The doctor's head lifted again and looked down at him.

It was a different doctor. This one was a far younger man, with an ascetic face and gleaming, almost fey eyes. There was no doubt about it. One of the last conscious thoughts he had had was that of gratitude that the attending physician, there at the end, had not been the one who secretly hated him for his one-time associations with the Nazi hierarchy. The attending doctor, instead, had worn an expression amusingly proper for that of a Swiss expert called to the deathbed of an eminent man: a mixture of worry at the prospect of losing so eminent a patient, and complacency at the thought that, at the old man's age, nobody could blame this doctor if

he died. At 85, pneumonia is a serious matter, with or
without penicillin.

"You're all right now," the new doctor said, freeing his
patient's head of a whole series of little silver rods which
had been clinging to it by a sort of network cap. "Rest a
minute and try to be calm. Do you know your name?"

He drew a cautious breath. There seemed to be nothing
at all the matter with his lungs now; indeed, he felt positively
healthy. "Certainly," he said, a little nettled. "Do you know
yours?"

The doctor smiled crookedly. "You're in character, it appears," he said. "My name is Barkun Kris; I am a mind
sculptor. Yours?"

"Richard Strauss."

"Very good," Dr. Kris said, and turned away. Strauss,
however, had already been diverted by a new singularity.
*Strauss* is a word as well as a name in German; it has many
meanings—an ostrich, a bouquet; von Wolzogen had had a
high old time working all the possible puns into the libretto
of *Feuersnot*. And it happened to be the first German word
to be spoken either by himself or by Dr. Kris since that
twice-removed moment of death. The language was not
French or Italian, either. It was most like English, but not
the English Strauss knew; nevertheless, he was having no
trouble speaking it and even thinking in it.

*Well*, he thought, *I'll be able to conduct* The Love of Danae
*after all. It isn't every composer who can première his own
opera posthumously*. Still, there was something queer about
all this— the queerest part of all being that conviction, which
would not go away, that he had actually been dead for just
a short time. Of course medicine was making great strides,
but . . .

"Explain all this," he said, lifting himself to one elbow.
The bed was different, too, and not nearly as comfortable as
the one in which he had died. As for the room, it looked
more like a dynamo shed than a sickroom. Had modern medicine taken to reviving its corpses on the floor of the Sie-
manns-Schukert plant?

"In a moment," Dr. Kris said. He finished rolling some
machine back into what Strauss impatiently supposed to be
its place, and crossed to the pallet. "Now. There are many
things you'll have to take for granted without attempting to
understand them, Dr. Strauss. Not everything in the world

today is explicable in terms of your assumptions. Please bear
that in mind."

"Very well. Proceed."

"The date," Dr. Kris said, "is 2161 by your calendar—
or, in other words, it is now two hundred and twelve years
after your death. Naturally, you'll realize that by this time
nothing remains of your body but the bones. The body you
have now was volunteered for your use. Before you look
into a mirror to see what it's like, remember that its physical
difference from the one you were used to is all in your
favor. It's in perfect health, not unpleasant for other people
to look at, and its physiological age is about fifty."

A miracle? No, not in this new age, surely. It was simply
a work of science. But what a science! This was Nietzsche's
eternal recurrence and the immortality of the superman
combined into one.

"And where is this?" the composer said.

"In Port York, part of the State of Manhattan, in the
United States. You will find the country less changed in
some respects than I imagine you anticipate. Other changes,
of course, will seem radical to you; but it's hard for me to
predict which ones will strike you that way. A certain
resilience on your part will bear cultivating."

"I understand," Strauss said, sitting up. "One question,
please; is it still possible for a composer to make a living in
this century?"

"Indeed it is," Dr. Kris said, smiling. "As we expect you
to do. It is one of the purposes for which we've—brought
you back."

"I gather, then," Strauss said somewhat dryly, "that there
is still a demand for my music. The critics in the old
days——"

"That's not quite how it is," Dr. Kris said. "I understand
some of your work is still played, but frankly I know very
little about your current status. My interest is rather——"

A door opened somewhere, and another man came in. He
was older and more ponderous than Kris and had a certain
air of academicism; but he too was wearing the oddly
tailored surgeon's gown, and looked upon Kris's patient
with the glowing eyes of an artist.

"A success, Kris?" he said. "Congratulations."

"They're not in order yet," Dr. Kris said. "The final
proof is what counts. Dr. Strauss, if you feel strong enough,

Dr. Seirds and I would like to ask you some questions. We'd like to make sure your memory is clear."

"Certainly. Go ahead."

"According to our records," Kris said, "you once knew a man whose initials were RKL; this was while you were conducting at the Vienna *Staatsoper*." He made the double "a" at least twice too long, as though German were a dead language he was striving to pronounce in some "classical" accent. "What was his name, and who was he?"

"That would be Kurt List—his first name was Richard, but he didn't use it. He was assistant stage manager."

The two doctors looked at each other. "Why did you offer to write a new overture to *The Woman Without a Shadow*, and give the manuscript to the City of Vienna?"

"So I wouldn't have to pay the garbage removal tax on the Maria Theresa villa they had given me."

"In the back yard of your house at Garmisch-Partenkirchen there was a tombstone. What was written on it?"

Strauss frowned. That was a question he would be happy to be unable to answer. If one is to play childish jokes upon oneself, it's best not to carve them in stone, and put the carving where you can't help seeing it every time you go out to tinker with the Mercedes. "It says," he replied wearily, "*Sacred to the memory of Guntram, Minnesinger, slain in a horrible way by his father's own symphony orchestra.*"

"When was *Guntram* premièred?"

"In—let me see—1894, I believe."

"Where?"

"In Weimar."

"Who was the leading lady?"

"Pauline de Ahna."

"What happened to her afterward?"

"I married her. Is she . . ." Strauss began anxiously.

"No," Dr. Kris said. "I'm sorry, but we lack the data to reconstruct more or less ordinary people."

The composer sighed. He did not know whether to be worried or not. He had loved Pauline, to be sure; on the other hand, it would be pleasant to be able to live the new life without being forced to take off one's shoes every time one entered the house, so as not to scratch the polished hardwood floors. And also pleasant, perhaps, to have two o'clock in the afternoon come by without hearing Pauline's everlasting, *"Richard—jetzt komponiert!"*

"Next question," he said.

For reasons which Strauss did not understand, but was content to take for granted, he was separated from Drs. Kris and Seirds as soon as both were satisfied that the composer's memory was reliable and his health stable. His estate, he was given to understand, had long since been broken up—a sorry end for what had been one of the principal fortunes of Europe—but he was given sufficient money to set up lodgings and resume an active life. He was provided, too, with introductions which proved valuable.

It took longer than he had expected to adjust to the changes that had taken place in music alone. Music was, he quickly began to suspect, a dying art, which would soon have a status not much above that held by flower arranging back in what he thought of as his own century. Certainly it couldn't be denied that the trend toward fragmentation, already visible back in his own time, had proceeded almost to completion in 2161.

He paid no more attention to American popular tunes than he had bothered to pay in his previous life. Yet it was evident that their assembly-line production methods—all the ballad composers openly used a slide-rule-like device called a Hit Machine—now had their counterparts almost throughout serious music.

The conservatives these days, for instance, were the twelve-tone composers—always, in Strauss's opinions, a dryly mechanical lot, but never more so than now. Their gods—Berg, Schoenberg, von Webern—were looked upon by the concert-going public as great masters, on the abstruse side perhaps, but as worthy of reverence as any of the Three B's.

There was one wing of the conservatives, however, which had gone the twelve-tone procedure one better. These men composed what was called "stochastic music," put together by choosing each individual note by consultation with tables of random numbers. Their bible, their basic text, was a volume called *Operational Aesthetics,* which in turn derived from a discipline called information theory; and not one word of it seemed to touch upon any of the techniques and customs of composition which Strauss knew. The ideal of this group was to produce music which would be "universal"—that is, wholly devoid of any trace of the composer's individuality, wholly a musical expression of the universal Laws of Chance. The Laws of Chance seemed to have a style of their own, all

right; but to Strauss it seemed the style of an idiot child
being taught to hammer a flat piano, to keep him from
getting into trouble.

By far the largest body of work being produced, however,
fell into a category misleadingly called "science-music." The
term reflected nothing but the titles of the works, which dealt
with space flight, time travel, and other subjects of a romantic
or an unlikely nature. There was nothing in the least sci-
entific about the music, which consisted of a mélange of
clichés and imitations of natural sounds, in which Strauss was
horrified to see his own time-distorted and diluted image.

The most popular form of science-music was a nine-minute
composition called a concerto, though it bore no re-
semblance at all to the classical concerto form; it was instead
a sort of free rhapsody after Rachmaninoff—long after. A
typical one—"Song of Deep Space" it was called, by somebody named H. Valerion Krafft—began with a loud assault on
the tam-tam, after which all the strings rushed up the scale in
unison, followed at a respectful distance by the harp and
one clarinet in parallel 6/4's. At the top of the scale cymbals
were bashed together, *forte possibile*, and the whole orchestra
launched itself into a major-minor, wailing sort of melody;
the whole orchestra, that is, except for the French horns,
which were plodding back down the scale again in what was
evidently supposed to be a countermelody. The second
phrase of the theme was picked up by a solo trumpet with
a suggestion of tremolo; the orchestra died back to its roots
to await the next cloudburst, and at this point—as any four-
year-old could have predicted—the piano entered with the
second theme.

Behind the orchestra stood a group of thirty women, ready
to come in with a wordless chorus intended to suggest the
eeriness of Deep Space—but at this point, too, Strauss had
already learned to get up and leave. Atfer a few such ex-
periences he could also count upon meeting in the lobby Sindi
Noniss, the agent to whom Dr. Kris had introduced him, and
who was handling the reborn composer's output—what there
was of it thus far. Sindi had come to expect these walkouts
on the part of his client, and patiently awaited them, standing beneath a bust of Gian Carlo Menotti; but he liked
them less and less, and lately had been greeting them by
turning alternately red and white like a toti-potent barber
pole.

"You shouldn't have done it," he burst out after the Krafft incident. "You can't just walk out on a new Krafft composition. The man's the president of the Interplanetary Society for Contemporary Music. How am I ever going to persuade them that you're a contemporary if you keep snubbing them?"

"What does it matter?" Strauss said. "They don't know me by sight."

"You're wrong; they know you very well, and they're watching every move you make. You're the first major composer the mind sculptors ever tackled, and the ISCM would be glad to turn you back with a rejection slip."

"Why?"

"Oh," said Sindi, "there are lots of reasons. The sculptors are snobs; so are the ISCM boys. Each of them wants to prove to the other that their own art is the king of them all. And then there's the competition; it would be easier to flunk you than to let you into the market. I really think you'd better go back in. I could make up some excuse——"

"No," Strauss said shortly. "I have work to do."

"But that's just the point, Richard. How are we going to get an opera produced without the ISCM? It isn't as though you wrote theremin solos, or something that didn't cost so——"

"I have work to do," he said, and left.

And he did: work which absorbed him as had no other project during the last thirty years of his former life. He had scarcely touched pen to music paper—both had been astonishingly hard to find—when he realized that nothing in his long career had provided him with touchstones by which to judge what music he should write *now*.

The old tricks came swarming back by the thousands, to be sure: the sudden, unexpected key changes at the crest of a melody; the interval stretching; the piling of divided strings, playing in the high harmonics, upon the already tottering top of a climax; the scurry and bustle as phrases were passed like lightning from one choir of the orchestra to another; the flashing runs in the brass, the chuckling in the clarinets, the snarling mixtures of colors to emphasize dramatic tension— all of them.

But none of them satisfied him now. He had been content with them for most of a lifetime, and had made them do an astonishing amount of work. But now it was time to strike out afresh. Some of the tricks, indeed, actively repelled him: where had he gotten the notion, clung to for decades, that violins screaming out in unison somewhere in the stratosphere was a sound interesting enough to be worth repeating inside a single composition, let alone in all of them?

And nobody, he reflected contentedly, ever approached such a new beginning better equipped. In addition to the past lying available in his memory, he had always had a technical armamentarium second to none; even the hostile critics had granted him that. Now that he was, in a sense, composing his first opera—his first after fifteen of them!—he had every opportunity to make it a masterpiece.

And every such intention.

There were, of course, many minor distractions. One of them was that search for old-fashioned score paper, and a pen and ink with which to write on it. Very few of the modern composers, it developed, wrote their music at all. A large bloc of them used tape, patching together snippets of tone and sound snipped from other tapes, superimposing one tape on another, and varying the results by twirling an elaborate array of knobs this way or that. Almost all the composers of 3-V scores, on the other hand, wrote on the sound track itself, rapidly scribbling jagged wiggly lines which, when passed through a photocell-audio circuit, produced a noise reasonably like an orchestra playing music, overtones and all.

The last-ditch conservatives who still wrote notes on paper, did so with the aid of a musical typewriter. The device, Strauss had to admit, seemed perfected at last; it had manuals and stops like an organ, but it was not much more than twice as large as a standard letter-writing typewriter, and produced a neat page. But he was satisfied with his own spidery, highly-legible manuscript and refused to abandon it, badly though the one pen nib he had been able to buy coarsened it. It helped to tie him to his past.

Joining the ISCM had also caused him some bad moments, even after Sindi had worked him around the political road blocks. The Society man who examined his qualifications as a member had run through the questions with no more interest than might have been shown by a veterinarian examining his four thousandth sick calf.

"Had anything published?"

"Yes, nine tone poems, about three hundred songs, an——"

"Not when you were alive," the examiner said, somewhat disquietingly. "I mean since the sculptors turned you out again."

"Since the sculptors—ah, I understand. Yes, a string quartet, two song cycles, a——"

"Good. Alfie, write down 'songs.' Play an instrument?"

"Piano."

"Hm." The examiner studied his fingernails. "Oh, well. Do you read music? Or do you use a Scriber, or tape clips? Or a Machine?"

"I read."

"Here." The examiner sat Strauss down in front of a viewing lectern, over the lit surface of which an endless belt of translucent paper was traveling. On the paper was an immensely magnified sound track. "Whistle me the tune of that, and name the instruments it sounds like."

"I don't read that *Musiksticheln*," Strauss said frostily, "or write it, either. I use standard notation, on music paper."

"Alfie, write down 'Reads notes only.' " He laid a sheet of grayly printed music on the lectern above the ground glass. "Whistle me that."

"That" proved to be a popular tune called "Vangs, Snifters and Store-Credit Snooky" which had been written on a Hit Machine in 2159 by a guitar-faking politician who sang it at campaign rallies. (In some respects, Strauss reflected, the United States had indeed not changed very much.) It had become so popular that anybody could have whistled it from the title alone, whether he could read the music or not. Strauss whistled it, and to prove his bona fides added, "It's in the key of B flat."

The examiner went over to the green-painted upright piano and hit one greasy black key. The instrument was horribly out of tune—the note was much nearer to the standard 440/cps A than it was to B flat—but the examiner said, "So it is. Alfie, write down, 'Also read flats.' All right, son, you're a member. Nice to have you with us; not many people can read that old-style notation any more. A lot of them think they're too good for it."

"Thank you," Strauss said.

"My feeling is, if it was good enough for the old masters, it's good enough for us. We don't have people like them with us these days, it seems to me. Except for Dr. Krafft, of course. They were *great* back in the old days—men like Shilkrit, Steiner, Tiomkin, and Pearl . . . and Wilder and Jannsen. Real goffin."

*"Doch gewiss,"* Strauss said politely.

But the work went forward. He was making a little income now, from small works. People seemed to feel a special interest in a composer who had come out of the mind sculptors' laboratories; and in addition the material itself, Strauss was quite certain, had merits of its own to help sell it.

It was the opera which counted, however. That grew and grew under his pen, as fresh and new as his new life, as founded in knowledge and ripeness as his long full memory. Finding a libretto had been troublesome at first. While it was possible that something existed that might have served among the current scripts for 3-V—though he doubted it— he found himself unable to tell the good from the bad through the fog cast over both by incomprehensibly technical production directions. Eventually, and for only the third time in his whole career, he had fallen back upon a play written in a language other than his own, and—for the first time—decided to set it in that language.

The play was Christopher Fry's *Venus Observed,* in all ways a perfect Strauss opera libretto, as he came gradually to realize. Though nominally a comedy, with a complex farcical plot, it was a verse play with considerable depth to it, and a number of characters who cried out to be brought by music into three dimensions, plus a strong undercurrent of autumnal tragedy, of leaf-fall and apple-fall—precisely the kind of contradictory dramatic mixture which von Hofmannsthal had supplied him with in *The Knight of the Rose,* in *Ariadne at Naxos,* and in *Arabella.*

Alas for von Hofmannsthal, but here was another long-dead playwright who seemed nearly as gifted; and the musical opportunities were immense. There was, for instance, the fire which ended act two; what a gift for a composer to whom orchestration and counterpoint were as important as air and water! Or take the moment where Perpetua shoots the apple from the Duke's hand; in that one moment a single passing reference could add Rossini's marmoreal *William Tell* to the musical texture as nothing but an ironic footnote! And the Duke's great curtain speech, beginning:

Shall I be sorry for myself? In Mortality's name

I'll be sorry for myself. Branches and boughs.
Brown hills, the valleys faint with brume,
A burnish on the lake . . .

*There* was a speech for a great tragic comedian, in the spirit of Falstaff; the final union of laughter and tears, punctuated by the sleepy comments of Reedbeck, to whose sonorous snore (trombones, no less than five of them, *con sordini?*) the opera would gently end. . . .

What could be better? And yet he had come upon the play only by the unlikeliest series of accidents. At first he had planned to do a straight knockabout farce, in the idiom of *The Silent Woman*, just to warm himself up. Remembering that Zweig had adapted that libretto for him, in the old days, from a play by Ben Jonson, Strauss had begun to search out English plays of the period just after Jonson's, and had promptly run aground on an awful specimen in heroic couplets called *Venice Preserv'd*, by one Thomas Otway. The Fry play had directly followed the Otway in the card catalogue, and he had looked at it out of curiosity; why should a Twentieth Century playwright be punning on a title from the Eighteenth?

After two pages of the Fry play, the minor puzzle of the pun disappeared entirely from his concern. His luck was running again; he had an opera.

Sindi worked miracles in arranging for the performance. The date of the première was set even before the score was finished, reminding Strauss pleasantly of those heady days when Fuerstner had been snatching the conclusion of *Elektra* off his work table a page at a time, before the ink was even dry, to rush it to the engraver before publication deadline. The situation now, however, was even more complicated, for some of the score had to be scribed, some of it taped, some of it engraved in the old way, to meet the new techniques of performance; there were moments when Sindi seemed to be turning quite gray.

But *Venus Observed* was, as usual, forthcoming complete from Strauss's pen in plenty of time. Writing the music in first draft had been hellishly hard work, much more like being reborn than had been that confused awakening in Barkun Kris's laboratory, with its overtones of being dead instead; but Strauss found that he still retained all of his old

ability to score from the draft almost effortlessly, as undis-
turbed by Sindi's half-audible worrying in the room with him
as he was by the terrifying supersonic bangs of the rockets
that bulleted invisibly over the city.

When he was finished, he had two days still to spare before
the beginning of rehearsals. With those, furthermore, he
would have nothing to do. The techniques of performance
in this age were so completely bound up with the electronic
arts as to reduce his own experience—he, the master *Kapell-
meister* of them all—to the hopelessly primitive.

He did not mind. The music, as written, would speak for
itself. In the meantime he found it grateful to forget the
months'-long preoccupation with the stage for a while. He
went back to the library and browsed lazily through old
poems, vaguely seeking texts for a song or two. He knew
better than to bother with recent poets; they could not speak
to him, and he knew it. The Americans of his own age, he
thought, might give him a clue to understanding this America
of 2161; and if some such poem gave birth to a song, so
much the better.

The search was relaxing and he gave himself up to enjoying it. Finally he struck a tape that he liked: a tape read
in a crackled old voice that twanged of Idaho as that voice had
twanged in 1910, in Strauss's own ancient youth. The poet's
name was Pound; he said, on the tape

> . . . the souls of all men great
> At times pass through us,
> And we are melted into them, and are not
> Save reflexions of their souls.
> Thus I am Dante for a space and am
> One François Villon, ballard-lord and thief
> Or am such holy ones I may not write,
> Lest Blasphemy be writ against my name;
> This for an instant and the flame is gone.
> 'Tis as in midmost us there glows a sphere
> Translucent, molten gold, that is the "I"
> And into this some form projects itself:
> Christus, or John, or eke the Florentine;
> And as the clear space is not if a form's
> Imposed thereon,
> So cease we from all being for the time,
> And these, the Masters of the Soul, live on.

He smiled. That lesson had been written again and again,

from Plato onward. Yet the poem was a history of his own case, a sort of theory for the metempsychosis he had undergone, and in its formal way it was moving. It would be fitting to make a little hymn of it, in honor of his own rebirth, and of the poet's insight.

A series of solemn, breathless chords framed themselves in his inner ear, against which the words might be intoned in a high, gently bending hush at the beginning . . . and then a dramatic passage in which the great names of Dante and Villon would enter ringing like challenges to Time. . . . He wrote for a while in his notebook before he returned the spool to its shelf.

These, he thought, are good auspices.

And so the night of the première arrived, the audience pouring into the hall, the 3-V cameras riding on no visible supports through the air, and Sindi calculating his share of his client's earnings by a complicated game he played on his fingers, the basic law of which seemed to be that one plus one equals ten. The hall filled to the roof with people from every class, as though what was to come would be a circus rather than an opera.

There were, surprisingly, nearly fifty of the aloof and aristocratic mind sculptors, clad in formal clothes which were exaggerated black versions of their surgeon's gowns. They had bought a bloc of seats near the front of the auditorium, where the gigantic 3-V figures which would shortly fill the "stage" before them (the real singers would perform on a small stage in the basement) could not but seem monstrously out of proportion; but Strauss supposed that they had taken this into account and dismissed it.

There was a tide of whispering in the audience as the sculptors began to trickle in, and with it an undercurrent of excitement the meaning of which was unknown to Strauss. He did not attempt to fathom it, however; he was coping with his own mounting tide of opening-night tension, which, despite all the years, he had never quite been able to shake.

The sourceless, gentle light in the auditorium dimmed, and Strauss mounted the podium. There was a score before him, but he doubted that he would need it. Directly before him, poking up from among the musicians, were the inevitable 3-V snouts, waiting to carry his image to the singers in the basement.

The audience was quiet now. This was the moment. His

baton swept up and then decisively down, and the prelude came surging up out of the pit.

For a little while he was deeply immersed in the always tricky business of keeping the enormous orchestra together and sensitive to the flexing of the musical web beneath his hand. As his control firmed and became secure, however, the task became slightly less demanding, and he was able to pay more attention to what the whole sounded like.

There was something decidedly wrong with it. Of course there were the occasional surprises as some bit of orchestral color emerged with a different *Klang* than he had expected; that happened to every composer, even after a lifetime of experience. And there were moments when the singers, entering upon a phrase more difficult to handle than he had calculated, sounded like someone about to fall off a tightrope (although none of them actually fluffed once; they were as fine a troupe of voices as he had ever had to work with).

But these were details. It was the over-all impression that was wrong. He was losing not only the excitement of the première—after all, that couldn't last at the same pitch all evening—but also his very interest in what was coming from the stage and the pit. He was gradually tiring; his baton arm becoming heavier; as the second act mounted to what should have been an impassioned outpouring of shining tone, he was so bored as to wish he could go back to his desk to work on that song.

Then the act was over; only one more to go. He scarcely heard the applause. The twenty minutes' rest in his dressing room was just barely enough to give him the necessary strength.

And suddenly, in the middle of the last act, he understood. There was nothing new about the music. It was the old Strauss all over again—but weaker, more dilute than ever. Compared with the output of composers like Krafft, it doubtless sounded like a masterpiece to this audience. But he knew.

The resolutions, the determination to abandon the old clichés and mannerisms, the decision to say something new—they had all come to nothing against the force of habit. Being brought to life again meant bringing to life as well all those deeply graven reflexes of his style. He had only to pick up his pen and they overpowered him with easy automatism, no more under his control than the jerk of a finger away from a flame.

His eyes filled; his body was young, but he was an old man, an old man. Another thirty-five years of this? Never. He had said all this before, centuries before. Nearly a half century condemned to saying it all over again, in a weaker and still weaker voice, aware that even this debased century would come to recognize in him only the burnt husk of greatness?—no; never, never.

He was aware, dully, that the opera was over. The audience was screaming its joy. He knew the sound. They had screamed that way when *Day of Peace* had been premièred, but they had been cheering the man he had been, not the man that *Day of Peace* showed with cruel clarity he had become. Here the sound was even more meaningless: cheers of ignorance, and that was all.

He turned slowly. With surprise, and with a surprising sense of relief, he saw that the cheers were not, after all, for him.

They were for Dr. Barkun Kris.

Kris was standing in the middle of the bloc of mind sculptors, bowing to the audience. The sculptors nearest him were shaking his hand one after the other. More grasped at it as he made his way to the aisle, and walked forward to the podium. When he mounted the rostrum and took the composer's limp hand, the cheering became delirious.

Kris lifted his arm. The cheering died instantly to an intent hush.

"Thank you," he said clearly. "Ladies and gentlemen, before we take leave of Dr. Strauss, let us again tell him what a privilege it has been for us to hear this fresh example of his mastery. I am sure no farewell could be more fitting."

The ovation lasted five minutes, and would have gone another five if Kris had not cut it off.

"Dr. Strauss," he said, "in a moment, when I speak a certain formulation to you, you will realize that your name is Jerom Bosch, born in our century and with a life in it all your own. The superimposed memories which have made you assume the mask, the *persona*, of a great composer will be gone. I tell you this so that you may understand why these people here share your applause with me."

A wave of assenting sound.

"The art of mind sculpture—the creation of artificial personalities for aesthetic enjoyment—may never reach such a pinnacle again. For you should understand that as Jerom Bosch you had no talent for music at all; indeed, we searched a long time to find a man who was utterly unable to carry even the simplest tune. Yet we were able to impose upon such unpromising material not only the personality, but the genius, of a great composer. That genius belongs entirely to you—to the *persona* that thinks of itself as Richard Strauss. None of the credit goes to the man who volunteered for the sculpture. That is your triumph, and we salute you for it."

Now the ovation could no longer be contained. Strauss, with a crooked smile, watched Dr. Kris bow. This mind sculpturing was a suitably sophisticated kind of cruelty for this age; but the impulse, of course, had always existed. It was the same impulse that had made Rembrandt and Leonardo turn cadavers into art works.

It deserved a suitably sophisticated payment under the *lex talionis:* an eye for an eye, a tooth for a tooth—and a failure for a failure.

No, he need not tell Dr. Kris that the "Strauss" he had created was as empty of genius as a hollow gourd. The joke would always be on the sculptor, who was incapable of hearing the hollowness of the music now preserved on the 3-V tapes.

But for an instant a surge of revolt poured through his blood stream. *I am I,* he thought. *I am Richard Strauss until I die, and will never be Jerom Bosch, who was utterly unable to carry even the simplest tune.* His hand, still holding the baton, came sharply up, though whether to deliver or to ward off a blow he could not tell.

He let it fall again, and instead, at last, bowed—not to the audience, but to Dr. Kris. He was sorry for nothing, as Kris turned to him to say the word that would plunge him back into oblivion, except that he would now have no chance to set that poem to music.

# To Pay the Piper

THE MAN in the white jacket stopped at the door marked *Re-Education Project—Col. H. H. Mudgett, Commanding Officer* and waited while the scanner looked him over. He had been through that door a thousand times, but the scanner made as elaborate a job of it as if it had never seen him before.

It always did, for there was always in fact a chance that it *had* never seen him before, whatever the fallible human beings to whom it reported might think. It went over him from gray, crew-cut poll to reagent-proof shoes, checking his small wiry body and lean profile against its stored silhouettes, tasting and smelling him as dubiously as if he were an orange held in storage two days too long.

"Name?" it said at last.

"Carson, Samuel, 32-454-0698."

"Business?"

"Medical director, Re-Ed One."

While Carson waited, a distant, heavy concussion came rolling down upon him through the mile of solid granite above his head. At the same moment, the letters on the door—and everything else inside his cone of vision—blurred distressingly, and a stab of pure pain went lancing through his head. It was the supersonic component of the explosion, and it was harmless—except that it always both hurt and scared him.

The light on the door-scanner, which had been glowing yellow up to now, flicked back to red again and the machine began the whole routine all over; the sound bomb had reset it. Carson patiently endured its inspection, gave his name, serial number, and mission once more, and this time got the green. He went in, unfolding as he walked the flimsy square of cheap paper he had been carrying all along.

Mudgett looked up from his desk and said at once: "What now?"

The physician tossed the square of paper down under Mudgett's eyes. "Summary of the press reaction to Hamelin's speech last night," he said. "The total effect is going against us, Colonel. Unless we can change Hamelin's mind, this outcry to re-educate civilians ahead of soldiers is going to lose the war for us. The urge to live on the surface again has been mounting for ten years; now it's got a target to focus on. Us."

Mudgett chewed on a pencil while he read the summary; a blocky, bulky man, as short as Carson and with hair as gray and close-cropped. A year ago, Carson would have told him that nobody in Re-Ed could afford to put stray objects in his mouth even once, let alone as a habit; now Carson just waited. There wasn't a man—or a woman or a child—of America's surviving thirty-five million "sane" people who didn't have some such tic. Not now, not after twenty-five years of underground life.

"He knows it's impossible, doesn't he?" Mudgett demanded abruptly.

"Of course he doesn't," Carson said impatiently. "He doesnt know any more about the real nature of the project than the people do. He thinks the 'educating' we do is in some sort of survival technique. . . . That's what the papers think, too, as you can plainly see by the way they loaded that editorial."

"Um. If we'd taken direct control of the papers in the first place . . ."

Carson said nothing. Military control of every facet of civilian life was a fact, and Mudgett knew it. He also knew that an appearance of freedom to think is a necessity for the human mind—and that the appearance could not be maintained without a few shreds of the actuality.

"Suppose we do this," Mudgett said at last. "Hamelin's position in the State Department makes it impossible for us to muzzle him. But it ought to be possible to explain to him that no unprotected human being can live on the surface, no matter how many Merit Badges he had for woodcraft and first aid. Maybe we could even take him on a little trip topside; I'll wager he's never seen it."

"And what if he dies up there?" Carson said stonily. "We lose three-fifths of every topside party as it is—and Hamelin's an inexperienced——"

"Might be the best thing, mightn't it?"

"*No,*" Carson said. "It would look like we'd planned it that way. The papers would have the populace boiling by the next morning."

Mudgett groaned and nibbled another double row of indentations around the barrel of the pencil. "There must be something," he said.

"There is."

"Well?"

"Bring the man here and show him just what we *are* doing. Re-educate *him,* if necessary. Once we told the newspapers that he'd taken the course . . . well, who knows, they just might resent it. Abusing his clearance privileges and so on."

"We'd be violating our basic policy," Mudgett said slowly. " 'Give the Earth back to the men who fight for it.' Still, the idea has some merits. . . ."

"Hamelin is out in the antechamber right now," Carson said. "Shall I bring him in?"

The radioactivity never did rise much beyond a mildly hazardous level, and that was only transient, during the second week of the war—the week called the Death of Cities. The small shards of sanity retained by the high commands on both sides dictated avoiding weapons with a built-in backfire; no cobalt bombs were dropped, no territories permanently poisoned. Generals still remembered that unoccupied territory, no matter how devastated, is still unconquered territory.

But no such considerations stood in the way of biological warfare. It was controllable: you never released against the enemy any disease you didn't yourself know how to control. There would be some slips, of course, but the margin for error . . .

There were some slips. But for the most part, biological warfare worked fine. The great fevers washed like tides around and around the globe, one after another. In such cities as had escaped the bombings, the rumble of truck convoys carrying the puffed heaped corpses to the mass graves became the only sound except for sporadic small-arms fire; and then that too ceased, and the trucks stood rusting in rows.

Nor were human beings the sole victims. Cattle fevers were sent out. Wheat rusts, rice molds, corn blights, hog choleras, poultry enteritises, fountained into the indifferent air from the hidden laboratories, or were loosed far aloft, in the jet-stream, by rocketing fleets. Gelatin capsules pullulating with gill-rots fell like hail into the great fishing grounds of Newfoundland, Oregon, Japan, Sweden, Portugal. Hundreds of species of animals were drafted as secondary hosts for human diseases, were injected and released to carry the blessings of the laboratories to their mates and litters. It was discovered that minute amounts of the tetracycline series of antibiotics, which had long been used as feed supplements to bring farm animals to full market weight early, could also be used to raise the most whopping Anopheles and Aëdes mosquitoes anybody ever saw, capable of flying long distances against the wind and of carrying a peculiarly interesting new strain of the malarial parasite and the yellow fever virus. . . .

By the time it had ended, everyone who remained alive was a mile under ground.

For good.

"I still fail to understand why," Hamelin said, "if, as you claim, you have methods of re-educating soldiers for surface life, you can't do so for civilians as well. Or instead."

The Under Secretary, a tall, spare man, bald on top, and with a heavily creased forehead, spoke with the odd neutral accent—untinged by regionalism—of the trained diplomat, despite the fact that there had been no such thing as a foreign service for nearly half a century.

"We're going to try to explain that to you," Carson said. "But we thought that, first of all, we'd try to explain once more why we think it would be bad policy—as well as physically out of the question.

"Sure, everybody wants to go topside as soon as it's possible. Even people who are reconciled to these endless caverns and corridors hope for something better for their children—a glimpse of sunlight, a little rain, the fall of a leaf. That's more important now to all of us than the war, which we don't believe in any longer. That doesn't even make any military sense, since we haven't the numerical strength to occupy the enemy's territory any more, and they haven't the strength to occupy ours. We understand all that. But we also know that the enemy is intent on prosecuting the war to the end. Extermination is what they say they want, on their propaganda broadcasts, and your own Department reports that they seem to mean what they say. So we can't give up fighting them; that would be simple suicide. Are you still with me?"

"Yes, but I don't see——"

"Give me a moment more. If we have to continue to fight, we know this much: that the first of the two sides to get men on the surface again—so as to be able to *attack* important targets, not just keep them isolated in seas of plagues—will be the side that will bring this war to an end. They know that, too. We have good reason to believe that they have a re-education project, and that it's about as far advanced as ours is."

"Look at it this way," Colonel Mudgett burst in unexpectedly. "What we have now is a stalemate. A saboteur occasionally locates one of the underground cities and lets the pestilences into it. Sometimes on our side, sometimes on theirs. But that only happens sporadically, and it's just more of this mutual extermination business—to which we're committed, willy-nilly, for as long as they are. If we can get troops onto the surface first, we'll be able to scout out their important installations in short order, and issue them a surrender ultimatum with teeth in it. They'll take it. The only other course is the sort of slow, mutual suicide we've got now."

Hamelin put the tips of his fingers together. "You gentlemen lecture me about policy as if I had never heard the word before. I'm familiar with your arguments for sending soldiers first. You assume that you're familiar with all of mine for starting with civilians, but you're wrong, because some of them haven't been brought up at all outside the Department. I'm going to tell you some of them, and I think they'll merit your close attention."

Carson shrugged. "I'd like nothing better than to be convinced, Mr. Secretary. Go ahead."

"You of all people should know, Dr. Carson, how close our underground society is to a psychotic break. To take a single instance, the number of juvenile gangs roaming these corridors of ours has increased 400 per cent since the rumors about the Re-Education Project began to spread. Or another: the number of individual crimes without motive—crimes committed just to distract the committer from the grinding monotony of the life we all lead—has now passed the total of all other crimes put together.

"And as for actual insanity—of our thirty-five million people still unhospitalized, there are four million cases *of which we know*, each one of which should be committed right now for early paranoid schizophrenia—except that were we to commit them, our essential industries would suffer a manpower loss more devastating than anything the enemy has inflicted upon us. Every one of those four million persons is a major hazard to his neighbors and to his job, but how can we do without them? And what can we do about the unrecognized, subclinical cases, which probably total twice as many? How long can we continue operating without a collapse under such conditions?"

Carson mopped his brow. "I didn't suspect that it had gone that far."

"It has gone that far," Hamelin said icily, "and it is accelerating. Your own project has helped to accelerate it. Colonel Mudgett here mentioned the opening of isolated cities to the pestilences. Shall I tell you how Louisville fell?"

"A spy again, I suppose," Mudgett said.

"No, Colonel. Not a spy. A band of—of vigilantes, of mutineers. I'm familiar with your slogan, 'The Earth to those who fight for it.' Do you know the counterslogan that's circulating among the people?"

They waited. Hamelin smiled and said: " 'Let's die on the surface.' "

"They overwhelmed the military detachment there, put the city administration to death, and blew open the shaft to the surface. About a thousand people actually made it to the top. Within twenty-four hours the city was dead—as the ringleaders had been warned would be the outcome. The warning didn't deter them. Nor did it protect the prudent citizens who had no part in the affair."

Hamelin leaned forward suddenly. "People won't wait to be told when it's their turn to be re-educated. They'll be tired of waiting, tired to the point of insanity of living at the bottom of a hole. They'll just go.

"And that, gentlemen, will leave the world to the enemy . . . or, more likely, the rats. They alone are immune to everything by now."

There was a long silence. At last Carson said mildly: "Why aren't *we* immune to everything by now?"

"Eh? Why—the new generations. They've never been exposed."

"We still have a reservoir of older people who lived through the war: people who had one or several of the new diseases that swept the world, some as many as five, and yet recovered. They still have their immunities. We know; we've tested them. We know from sampling that no new disease has been introduced by either side in over ten years now. Against all the known ones, we have immunization techniques, anti-sera, antibiotics, and so on. I suppose you get your shots every six months like all the rest of us; we should all be very hard to infect now, and such infections as do take should run mild courses." Carson held the Under Secretary's eyes grimly. "Now, answer me this question: why is it that, despite all these protections, *every single person* in an opened city dies?"

"I don't know," Hamelin said, staring at each of them in turn. "By your showing some of them should recover."

"They should," Carson said. "But nobody does. Why? Because the very nature of disease has changed since we all went underground. There are now abroad in the world a number of mutated bacterial strains which can by-pass the immunity mechanisms of the human body altogether. What this means in simple terms is that, should such a germ get into your body, your body wouldn't recognize it as an invader. It would manufacture no antibodies against the germ. Consequently, the germ could multiply without any check, and—you would die. So would we all."

"I see," Hamelin said. He seemed to have recovered his composure extraordinarily rapidly. "I am no scientist, gentlemen, but what you tell me makes our position sound perfectly hopeless. Yet obviously you have some answer."

Carson nodded. "We do. But it's important for you to understand the situation, otherwise the answer will mean nothing to you. So: is it perfectly clear to you now, from what we've said so far, that no amount of re-educating a man's brain, be he soldier *or* civilian, will allow him to survive on the surface?"

"Quite clear," Hamelin said, apparently ungrudgingly. Carson's hopes rose by a fraction of a millimeter. "But if you don't re-educate his brain, what can you re-educate? His reflexes, perhaps?"

"No," Carson said. "His lymph nodes, and his spleen."

A scornful grin began to appear on Hamelin's thin lips. "You need better public relations counsel than you've been getting," he said. "If what you say is true—as of course I assume it is—then the term 're-educate' is not only inappropriate, it's downright misleading. If you had chosen a less suggestive and more accurate label in the beginning, I wouldn't have been able to cause you half the trouble I have."

"I agree that we were badly advised there," Carson said. "But not entirely for those reasons. Of course the name is misleading; that's both a characteristic and a function of the names of top secret projects. But in this instance the name 'Re-Education,' bad as it now appears, subjected the men who chose it to a fatal temptation. You see, though it is misleading, it is also entirely accurate."

"Word games," Hamelin said.

"Not at all," Mudgett interposed. "We were going to spare you the theoretical reasoning behind our project, Mr. Secretary, but now you'll just have to sit still for it. The fact is that the body's ability to distinguish between its own cells and those of some foreign tissue—a skin graft, say, or a bacterial invasion of the blood—isn't an inherited ability. It's a learned reaction. Furthermore, if you'll think about it a moment, you'll see that it has to be. Body cells die, too, and have to be disposed of; what would happen if removing those dead cells provoked an antibody reaction, as the destruction of foreign cells does? We'd die of anaphylactic shock while we were still infants.

"For that reason, the body has to learn how to scavenge selectively. In human beings, that lesson isn't learned completely until about a month after birth. During the intervening time, the newborn infant is protected by antibodies that it gets from the colestrum, the 'first milk' it gets from the breast during the three or four days immediately after birth. It can't generate its own; it isn't allowed to, so to speak, until it's learned the trick of cleaning up body residues *without* triggering the antibody mechanisms. Any dead cells marked 'personal' have to be dealt with some other way."

"That seems clear enough," Hamelin said. "But I don't see its relevance."

"Well, we're in a position now where that differentiation between the self and everything outside the body doesn't do us any good any more. These mutated bacteria have been 'selfed' by the mutation. In other words, some of their protein molecules, probably desoxyribonucleic acid molecules, carry configurations or 'recognition units' identical with those of our body cells, so that the body can't tell one from another."

"But what has all this to do with re-education?"

"Just this," Carson said. "What we do here is to impose upon the cells of the body—all of them—a new set of recognition units for the guidance of the lymph nodes and the spleen, which are the organs that produce antibodies. The new units are highly complex, and the chances of their being duplicated by bacterial evolution, even under forced draft, are too small to worry about. That's what Re-Education is. In a few moments, if you like, we'll show you just how it's done."

Hamelin ground out his fifth cigarette in Mudgett's ash tray and placed the tips of his fingers together thoughtfully. Carson wondered just how much of the concept of recognition-marking the Under Secretary had absorbed. It had to be admitted that he was astonishingly quick to take hold of abstract ideas, but the self-marker theory of immunity was—like everything else in immunology—almost impossible to explain to laymen, no matter how intelligent.

"This process," Hamelin said hesitantly, "it takes a long time?"

"About six hours per subject, and we can handle only one man at a time. That means that we can count on putting no more than seven thousand troops into the field by the turn of the century. Every one will have to be a highly trained specialist, if we're to bring the war to a quick conclusion."

"Which means no civilians," Hamelin said. "I see. I'm not entirely convinced, but—by all means let's see how it's done."

Once inside, the Under Secretary tried his best to look everywhere at once. The room cut into the rock was roughly two hundred feet high. Most of it was occupied by the bulk of the Re-Education Monitor, a mechanism as tall as a fifteen-story building, and about a city block square. Guards watched it on all sides, and the face of the machine swarmed with technicians.

"Incredible," Hamelin murmured. "That enormous object can process only one man at a time?"

"That's right," Mudgett said. "Luckily it doesn't have to treat all the body cells directly. It works through the blood, re-selfing the cells by means of small changes in the serum chemistry."

"What kind of changes?"

"Well," Carson said, choosing each word carefully, "that's more or less a graveyard secret, Mr. Secretary. We can tell you this much: the machine uses a vast array of crystalline, complex sugars which *behave* rather like the blood-group-and-type proteins. They're fed into the serum in minute amounts, under feedback control of second-by-second analysis of the blood. The computations involved in deciding upon the amount and the precise nature of each introduced chemical are highly complex. Hence the size of the machine. It is, in its major effect, an artificial kidney."

"I've seen artificial kidneys in the hospitals," Hamelin said, frowning. "They're rather compact affairs."

"Because all they do is remove waste products from the patient's blood, and restore the fluid and electrolyte balance. Those are very minor renal functions in the higher mammals. The organ's main duty is chemical control of immunity. If Burnet and Fenner had known that back in 1949, when the selfing theory was being formulated, we'd have had Re-Education long before now."

"Most of the machine's size is due to the computation section," Mudgett emphasized. "In the body, the brain stem does those computations, as part of maintaining homeostasis. But we can't reach the brain stem from outside; it's not under conscious control. Once the body is re-selfed, it will retrain the thalamus where we can't." Suddenly, two swinging doors at the base of the machine were pushed apart and a mobile operating table came through, guided by two attendants. There was a form on it, covered to the chin with a sheet. The face above this sheet was immobile and almost as white.

Hamelin watched the table go out of the huge cavern with visibly mixed emotions. He said: "This process—it's painful?"

"No, not exactly," Carson said. The motive behind the question interested him hugely, but he didn't dare show it. "But any fooling around with the immunity mechanisms can give rise to symptoms—fever, general malaise, and so on. We try to protect our subjects by giving them a light shock anesthesia first."

"Shock?" Hamelin repeated. "You mean electroshock? I don't see how——"

"Call it stress anesthesia instead. We give the man a steroid drug that counterfeits the anesthesia the body itself produces in moments of great stress—on the battlefield, say, or just after a serious injury. It's fast, and free of aftereffects. There's no secret about that, by the way; the drug involved is 21-hydroxypregnane-3,20-dione sodium succinate, and it dates all the way back to 1955."

"Oh," the Under Secretary said. The ringing sound of the chemical name had had, as Carson had hoped, a ritually soothing effect.

"Gentlemen," Hamelin said hesitantly. "Gentlemen, I have a—a rather unusual request. And, I am afraid, a rather selfish one." A brief, nervous laugh. "Selfish in both senses, if you will pardon me the pun. You need feel no hesitation in refusing me, but——"

Abruptly he appeared to find it impossible to go on. Carson mentally crossed his fingers and plunged in.

"You would like to undergo the process yourself?" he said.

"Well, yes. Yes, that's exactly it. Does that seem inconsistent? I should know, should I not, what it is that I'm advocating for my following? Know it intimately, from personal experience, not just theory? Of course I realize that it would conflict with your policy, but I assure you I wouldn't turn it to any political advantage—none whatsoever. And perhaps it wouldn't be too great a lapse of policy to process just one civilian among your seven thousand soldiers."

Subverted, by God! Carson looked at Mudgett with a firmly straight face. It wouldn't do accept too quickly.

But Hamelin was rushing on, almost chattering now. "I

can understand your hesitation. You must feel that I'm trying to gain some advantage, or even to get to the surface ahead of my fellow men. If it will set your minds at rest, I would be glad to enlist in your advance army. Before five years are up, I could surely learn some technical skill which would make me useful to the expedition. If you would prepare papers to that effect, I'd be happy to sign them."

"That's hardly necessary," Mudgett said. "After you're Re-Educated, we can simply announce the fact, and say that you've agreed to join the advance party when the time comes."

"Ah," Hamelin said. "I see the difficulty. No, that would make my position quite impossible. If there is no other way . . ."

"Excuse us a moment," Carson said. Hamelin bowed, and the doctor pulled Mudgett off out of earshot.

"Don't overplay it," he murmured. "You're tipping our hand with that talk about a press release, Colonel. He's offering us a bribe—but he's plenty smart enough to see that the price you're suggesting is that of his whole political career; he won't pay that much."

"What then?" Mudgett whispered hoarsely.

"Get somebody to prepare the kind of informal contract he suggested. Offer to put it under security seal so we won't be able to show it to the press at all. He'll know well enough that such a seal can be broken if our policy ever comes before a presidential review—and that will restrain him from forcing such a review. Let's not demand too much. Once he's been Re-Educated, he'll have to live the rest of the five years with the knowledge that he *can* live topside any time he wants to try it—and he hasn't had the discipline our men have had. It's my bet that he'll goof off before the five years are up—and good riddance."

They went back to Hamelin, who was watching the machine and humming in a painfully abstracted manner.

"I've convinced the Colonel," Carson said, "that your services in the army might well be very valuable when the time comes, Mr. Secretary. If you'll sign up, we'll put the papers under security seal for your own protection, and then I think we can fit you into our treatment program today."

"I'm grateful to you, Dr. Carson," Hamelin said. "Very grateful indeed."

Five minutes after his injection, Hamelin was as peaceful as a flounder and was rolled through the swinging doors. An hour's discussion of the probable outcome, carried on in the privacy of Mudgett's office, bore very little additional fruit, however.

"It's our only course," Carson said. "It's what we hoped to gain from his visit, duly modified by circumstances. It all comes down to this: Hamelin's compromised himself, and he knows it."

"But," Mudgett said, "suppose he was right? What about all that talk of his about mass insanity?"

"I'm sure it's true," Carson said, his voice trembling slightly despite his best efforts at control. "It's going to be rougher than ever down here for the next five years, Colonel. Our only consolation is that the enemy must have exactly the same problem; and if we can beat them to the surface——"

"*Hsst!*" Mudgett said. Carson had already broken off his sentence. He wondered why the scanner gave a man such a hard time outside that door, and then admitted him without any warning to the people on the other side. Couldn't the damned thing be trained to knock?

The newcomer was a page from the haemotology section. "Here's the preliminary rundown on your 'student X,' Dr. Carson," he said.

The page saluted Mudgett and went out. Carson began to read. After a moment, he also began to sweat.

"Colonel, look at this. I was wrong after all. Disastrously wrong. I haven't seen a blood-type distribution pattern like Hamelin's since I was a medical student, and even back then it was only a demonstration, not a real live patient. Look at it from the genetic point of view—the migration factors."

He passed the protocol across the desk. Mudgett was not by background a scientist, but he was an enormously able administrator, of the breed that makes it its business to know the technicalities on which any project ultimately rests. He was not much more than halfway through the tally before his eyebrows were gaining altitude like shock waves.

"Carson, we can't let that man into the machine! He's——"

"He's already in it, Colonel, you know that. And if we interrupt the process before it runs to term, we'll kill him."

"Let's kill him, then," Mudgett said harshly. "Say he died while being processed. Do the country a favor."

"That would produce a hell of a stink. Besides, we have no proof."

Mudgett flourished the protocol excitedly.

"That's not proof to anyone but a haemotologist."

"But Carson, the man's a saboteur!" Mudgett shouted. "Nobody but an Asiatic could have a typing pattern like this! And he's no melting-pot product, either—he's a classical mixture, very probably a Georgian. And every move he's made since we first heard of him has been aimed directly at us—aimed directly at tricking us into getting him into the machine!"

"I think so too," Carson said grimly. "I just hope the enemy hasn't many more agents as brilliant."

"One's enough," Mudgett said. "He's sure to be loaded to the last cc of his blood with catalyst poisons. Once the machine starts processing his serum, we're done for—it'll take us years to reprogram the computer, if it can be done at all. It's *got* to be stopped!"

"Stopped?" Carson said, astonished. "But it's already stopped. That's not what worries me. The machine stopped it fifty minutes ago."

"It can't have! How could it? It has no relevant data!"

"Sure it has." Carson leaned forward, took the cruelly chewed pencil away from Mudgett, and made a neat check beside one of the entries on the protocol. Mudgett stared at the checked item.

"Platelets Rh VI?" he mumbled. "But what's that got to do with . . . Oh. Oh, I see. That platelet type doesn't exist at all in our population now, does it? Never seen it before myself, at least."

"No," Carson said, grinning wolfishly. "It never was common in the West, and the pogrom of 1981 wiped it out. That's something the enemy couldn't know. But the machine knows it. As soon as it gives him the standard anti-IV desensitization shot, his platelets will begin to dissolve—and he'll be rejected for incipient thrombocytopenia." He laughed. "For his own protection! But——"

"But he's getting nitrous oxide in the machine, and he'll be held six hours under anesthesia anyhow—also for his own protection," Mudgett broke in. He was grinning back at Carson like an idiot. "When he comes out from under, he'll assume that he's been re-educated, and he'll beat it back to the enemy to report that he's poisoned our machine, so that they can be sure they'll beat us to the surface. And he'll go the fastest way: *overland.*"

"He will," Carson agreed. "Of course he'll go overland, and of course he'll die. But where does that leave us? We won't be able to conceal that he was treated here, if there's any sort of inquiry at all. And his death will make everything we do here look like a fraud. Instead of paying our Pied Piper—and great jumping Jehoshaphat, look at his name! They were rubbing our noses in it all the time! Nevertheless, we didn't pay the piper; we killed him. And 'platelets Rh VI' won't be an adequate excuse for the press, or for Hamelin's following."

"It doesn't worry me," Mudgett rumbled. "Who'll know? He won't die in our labs. He'll leave here hale and hearty. He won't die until he makes a break for the surface. After that we can compose a fine obituary for the press. Heroic government official, on the highest policy level—couldn't wait to lead his followers to the surface—died of being too much in a hurry—Re-Ed Project sorrowfully reminds everyone that no technique is foolproof . . ."

Mudgett paused long enough to light a cigarette, which was a most singular action for a man who never smoked. "As a matter of fact, Carson," he said, "it's a natural."

Carson considered it. It seemed to hold up. And "Hamelin" would have a death certificate as complex as he deserved—not officially, of course, but in the minds of everyone who knew the facts. His death, when it came, would be due directly to the thrombocytopenia which had caused the Re-Ed machine to reject him—and thrombocytopenia is a disease of infants. *Unless ye become as little children . . .*

That was a fitting reason for rejection from the new kingdom of Earth: anemia of the newborn.

His pent breath went out of him in a long sigh. He hadn't been aware that he'd been holding it. "It's true," he said softly. "That's the time to pay the piper."

"When?" Mudget said.

"When?" Carson said, surprised. "Why, *before* he takes the children away."

## Nor Iron Bars

THE *Flyaway II*, which was large enough to carry a hundred passengers, seemed twice as large to Gordon Arpe with only the crew on board—large and silent, with the silence of its orbit a thousand miles above the Earth.

"When are they due?" Dr. (now Captain) Arpe said, for at least the fourth time. His second officer, Friedrich Oestreicher, looked at the chronometer and away again with boredom.

"The first batch will be on board in five minutes," he said harshly. "Presumably they've all reached SV-One by now. It only remains to ferry them over."

Arpe nibbled at a fingernail. Although he had always been the tall, thin, and jumpy type, nail-biting was a new vice to him.

"I still think it's insane to be carrying passengers on a flight like this," he said.

Oestreicher said nothing. Carrying passengers was no novelty to him. He had been captain of a passenger vessel on the Mars run for ten years, and looked it: a stocky hard-muscled youngster of thirty, whose crew cut was going gray despite the fact that he was five years younger than Arpe. He was second in command of the *Flyaway II* only because he had no knowledge of the new drive. Or, to put it another way, Arpe was captain only because he was the only man who did understand it, having invented it. Either way you put it didn't sweeten it for Oestreicher, that much was evident.

Well, the first officer would be the acting captain most of the time, anyhow. Arpe admitted that he himself had no knowledge of how to run a space ship. The thought of passengers, furthermore, came close to terrifying him. He hoped to have as little contact with them as possible.

But dammitall, it *was* crazy to be carrying a hundred laymen—half of them women and children, furthermore—on the maiden flight of an untried interstellar drive, solely on the belief of one Dr. Gordon Arpe that his brain child would work. Well, that wasn't the sole reason, of course. The whole Flyaway project, of which Arpe had been head, believed it would work, and so did the government.

And then there was the First Expedition to Centaurus, presumably still in flight after twelve years; they had elected to do it the hard way, on ion drive, despite Garrard's spectacular solo round trip, the Haertel overdrive which had made that possible being adjudged likely to be damaging to the sanity of a large crew. Arpe's discovery had been a totally unexpected breakthrough, offering the opportunity to rush a new batch of trained specialists to help the First Expedition colonize, arriving only a month or so after the First had landed. And if you are sending help, why not send families, too—the families the First Expedition had left behind?

Which also explained the two crews. One of them consisted of men from the Flyaway project, men who had built various parts of the drive, or designed them, or otherwise knew them intimately. The other was made up of men who had served some time—in some cases, as long as two full hitches—in the Space Service under Oestreicher. There was some overlapping, of course. The energy that powered the drive field came from a Nernst-effect generator: a compact ball of fusing hydrogen, held together in mid-combustion chamber by a hard magnetic field, which transformed the heat into electricity to be bled off perpendicular to the magnetic lines of force. The same generator powered the ion rockets of ordinary interplanetary flight, and so could be serviced by ordinary crews. On the other hand, Arpe's new attempt to beat the Lorenz-Fitzgerald equation involved giving the whole ship negative mass, a concept utterly foreign to even the most experienced spaceman. Only a physicist who knew Dirac holes well enough to call them "Pam" would have thought of the notion at all.

But it would work. Arpe was sure of that. A body with negative mass could come very close to the speed of light before the Fitzgerald contraction caught up with it, and without the wild sine-curve variation in subjective time which the non-Fitzgeraldian Haertel overdrive enforced on the passenger. If the field could be maintained successfully in spite of the contraction, there was no good reason why the velocity of light could not be passed; under such conditions, the ship would not be a material object at all.

And polarity in mass does not behave like polarity in electromagnetic fields. As gravity shows, where mass is concerned like attracts like, and unlikes repel. The very charging of the field should fling the charged object away from the Earth at a considerable speed.

The unmanned models had not been disappointing. They had vanished instantly, with a noise like a thunderclap. And since every atom in the ship was affected evenly, there ought to be no sensible acceleration, either—which is a primary requirement for an ideal drive. It looked good . . .

But not for a first test with a hundred passengers!

"Here they come," said Harold Stauffer, the second officer. Sandy-haired and wiry, he was even younger than Oestreicher, and had the small chin combined with handsome features which is usually called "a weak face." He was, Arpe already knew, about as weak as a Diesel locomotive; so much for physiognomy. He was pointing out the viewplate.

Arpe started and followed the pointing finger. At first he saw nothing but the doughnut with the peg in the middle which was Satellite Vehicle 1, as small as a fifty-cent piece at this distance. Then a tiny sliver of flame near it disclosed the first of the ferries, coming toward them.

"We had better get down to the air lock," Oestreicher said.

"All right," Arpe responded abstractedly. "Go ahead. I still have some checking to do."

"Better delegate it," Oestreicher said. "It's traditional for the captain to meet passengers coming on board. They expect it. And this batch is probably pretty scared, considering what they've undertaken. I wouldn't depart from routine with them if I were you, sir."

"I can run the check," Stauffer said helpfully. "If I get into any trouble on the drive, sir, I can always call your gang chief. He can be the judge of whether or not to call you."

Outgeneraled, Arpe followed Oestreicher down to the air lock.

The first ferry stuck its snub nose into the receiving area; the nose promptly unscrewed and tipped upward. The first passenger out was a staggering two-year-old, as bundled up as though it had been dressed for "the cold of space," so that nobody could have told whether it was a boy or a girl. It fell down promptly, got up again without noticing, and went charging straight ahead, shouting "Bye-bye-see-you, bye-bye-see-you, bye-bye——" Then it stopped, transfixed, looking about the huge metal cave with round eyes.

"Judy?" a voice cried from inside the ferry. "Judy! Judy, wait for Mommy!"

After a moment, the voice's owner emerged: a short, fair girl, perhaps eighteen. The baby by this time had spotted the crew member who had the broadest grin, and charged him shouting "Daddy Daddy Daddy Daddy Daddy Daddy" like a machine gun. The woman followed, blushing.

The crewman was not embarrassed. It was obvious that he had been called Daddy before by infants on three planets and five satellites, with what accuracy he might not have been able to guarantee. He picked up the little girl and poked her gently.

"Hi-hi, Judy," he said. "I see you. Where's Judy? *I* see her."

Judy crowed and covered her face with her hands; but she was peeking.

"Something's wrong here," Arpe murmured to Oestreicher. "How can a man who's been traveling toward Centaurus for twelve years have a two-year-old daughter?"

"Wouldn't raise the question if I were you, sir," Oestreicher said through motionless lips. "Passengers are never a uniform lot. Best to get used to it."

The aphorism was being amply illustrated. Next to leave the ferry was an old woman who might possibly have been the mother of one of the crewmen of the First Centaurus expedition; by ordinary standards she was in no shape to stand a trip through space, and surely she would be no help to anybody when she arrived. She was followed by a striking brunette girl in close-fitting, close-cut leotards, with a figure like a dancer. She might have been anywhere between 21 and 41 years old; she wore no ring, and the hard set of her otherwise lovely face did not suggest that she was anybody's wife. Oddly, she also looked familiar. Arpe nudged Oestreicher and nodded toward her.

"Celia Gospardi," Oestreicher said out of the corner of his mouth. "Three-V comedienne. You've seen her, sir, I'm sure."

And so he had; but he would never have recognized her, for she was not smiling. Her presence here defied any explanation he could imagine.

"Screened or not, there's something irregular about this," Arpe said in a low voice. "Obviously there's been a slip in the interviewing. Maybe we can turn some of this lot back."

Oestreicher shrugged. "It's your ship, sir," he said. "I advise against it, however."

Arpe scarcely heard him. If some of these passengers were really as unqualified as they looked . . . and there would be no time to send up replacements . . . At random, he started with the little girl's mother.

"Excuse me, ma'am . . ."

The girl turned with surprise, and then with pleasure. "Yes, Captain!"

"Uh, it occurs to me that there may have been, uh, an error. The *Flyaway II's* passengers are strictly restricted to technical colonists and to, uh, legal relatives of the First Centaurus Expedition. Since your Judy looks to be no more than two, and since it's been twelve years since . . ."

The girl's eyes had already turned ice-blue; she rescued him, after a fashion, from a speech he had suddenly realized he could never have finished. "Judy," she said levelly, "is the granddaughter of Captain Willoughby of the First Expedition. I am his daughter. I am sorry my husband isn't alive to pin your ears back, Captain. Any further questions?"

Arpe left the field without stopping to collect his wounded. He was stopped in mid-retreat by a thirteen-year-old boy wearing astonishingly thick glasses and a thatch of hair that went in all directions in dirty blond cirri.

"Sir," the boy said, "I understood that this was to be a new kind of ship. It looks like an SC-Forty-seven freighter to me. Isn't it?"

"Yes," Arpe said. "Yes, that's what it is. That is, it's the same hull. I mean, the engines and fittings are new."

"*Uh*-huh," the boy said. He turned his back and resumed prowling.

The noise was growing louder as the reception area filled. Arpe was uncomfortably aware that Oestreicher was watching him with something virtually indistinguishable from contempt, but still he could not get away; a small, compact man in a gray suit had hold of his elbow.

"Captain Arpe, I'm Forrest of the President's Commission, to disembark before departure," he said in a low murmur, so rapidly that one syllable could hardly be told from another. "We've checked you out and you seem to be in good shape. Just want to remind you that your drive is more important than anything else on board. Get the passengers where they want to go by all means if it's feasible, but if it isn't, *the government wants that drive back*. That means jettisoning the passengers without compunction if necessary. Dig?"

"All right." That had been pounded into him almost from the beginning of his commission, but suddenly it didn't seem to be as clear-cut a proposition, not now, not after the passengers were actually arriving in the flesh. Filled with a sudden, unticketable emotion, almost like horror, Arpe shook the government man off. Bidding tradition be damned, he got back to the bridge as fast as he could go, leaving Oestreicher to cope with the remaining newcomers. After all, Oestreicher was supposed to know how.

But the rest of the ordeal still loomed ahead of him. The ship could not actually take off until "tomorrow," after a twelve-hour period during which the passengers would get used to their quarters, and got enough questions answered to prevent their wandering into restricted areas of the ship.

And there was still the traditional Captain's Dinner to be faced up to: a necessary ceremony during which the passengers got used to eating in free fall, got rid of their first awkwardness with the tools of space, and got to know each other, with the officers to help them. It was an initial step rather than a final one, as was the Captain's Dinner on the seas.

"Stauffer, how did the check-out go?"

"Mr. Stauffer, please, sir," the second officer said politely. "All tight, sir. I asked your gang chief to sign the log with me, which he did."

"Very good. Thank you—uh, Mr. Stauffer. Carry on."

"Yes, sir."

It looked like a long evening. Maybe Oestreicher would be willing to forgo the Captain's Dinner. Somehow, Arpe doubted that he would.

He wasn't willing, of course. He had already arranged for it long ago. Since there was no salon on the converted freighter, the dinner was held in one of the smaller holds, whose cargo had been strapped temporarily in the corridors. The whole inner surface of the hold was taken up by the saddle-shaped tables, to which the guests hitched themselves by belt hooks; service arrived from way up in the middle of the air.

Arpe's table was populated by the thirteen-year-old boy he had met earlier, a ship's nurse, two technicians from the specialists among the colonist-passengers, a Nernst-generator officer, and Celia Gospardi, who sat next to him. Since she had no children of her own with her, she had not been placed at one of the tables allocated to children and parents; besides, she was a celebrity.

Arpe was appalled to discover that she was not the only celebrity on board. At the very next table down was Daryon Hammersmith, the man the newscasts called "The Conqueror of Titan." There was no mistaking the huge-shouldered, flamboyant explorer and his heavy voice; he was a natural center of attention, especially among the women. He was bald, but this simply made him look even more like a Prussian officer of the old school, and as overpoweringly, cruelly masculine as a hunting panther.

For several courses Arpe could think of nothing at all to say. He rather hoped that this blankness of mind would last; maybe the passengers would gather that he was aloof by nature, and . . . But the silence at the captain's table was becoming noticeable, especially against the noise the children were making elsewhere. Next door, Hammersmith appeared to be telling stories.

And what stories! Arpe knew very little about the satellites, but he was somehow quite sure that there were no snow tigers on Titan who gnawed away the foundations of buildings, nor any three-eyed natives who relished frozen man-meat warmed just until its fluids changed from Ice IV to Ice III. If there were, it was odd that Hammersmith's own book about the Titan expedition had mentioned neither. But the explorer was making Arpe's silence even more conspicuous; he *had* to say something.

"Miss Gospardi—we're honored to have you with us. You have a husband among the First Expedition, I suppose?"

"Yes, worse luck," she said, gnawing with even white teeth at a drumstick. "My fifth."

"Oh. Well, if at first you don't succeed—isn't that how it goes? You're undertaking quite a journey to be with him again. I'm glad you feel so certain now."

"I'm certain," she said calmly. "It's a long trip, all right. But he made a big mistake when he thought it'd be too long for *me*."

The thirteen-year-old was watching her like an owl. It looked like a humid night for him.

"Of course, Titan's been tamed down considerably since my time," Hammersmith was booming jovially. "I'm told the new dome there is almost cozy, except for the wind. That wind—I still dream about it now and then."

"I admire your courage," Arpe said to the 3-V star, beginning to feel faintly courtly. Maybe he had talents he had neglected; he seemed to be doing rather well so far.

"It isn't courage," the woman said, freeing a piece of bread from the clutches of the Lazy Spider. "It's desperation. I hate space flight. I should know, I've had to make that Moon circuit for show dates often enough. But I'm going to get that lousy coward back if it's the last thing I do."

She took a full third out of the bread slice in one precise, gargantuan nibble.

"I wouldn't have thought of it if I hadn't lost my sixth husband to Peggy Walton. That skirt-chaser; I must have been out of my mind. But Johnny didn't bother to divorce me before he ran off on this Centaurus safari. That was a mistake. I'm going to haul *him* back by his *scruff*."

She folded the rest of the bread and snapped it delicately in two. The thirteen-year-old winced and looked away.

"No, I can't say that I miss Titan much," Hammersmith said, in a meditative tone which nevertheless carried the entire length of the hold. "I like planets where the sky is clear most of the time. My hobby is microastronomy—as a matter of fact I have some small reputation in the field, strictly as an amateur. I understand the stars should be unusually clear and brilliant in the Centaurus area, but of course there's nothing like open space for really serious work."

"To tell the truth," Celia went on, although for Arpe's money she had told more than enough truth already, "I'm scared to death of this bloated coffin of yours. But what the hell, I'm dead anyhow. On Earth, everybody knows I can't stay married two years, no matter how many fan letters I get. Or how many proposals, honorable or natural. It's no good to me any more that three million men say they love me. I know what they mean. Every time I take one of them up on it, he vanishes."

The folded snippet of bread vanished without a sound.

"Are you really going to be a colonist?" someone asked Hammersmith.

"Not for a while, anyhow," the explorer said. "I'm taking my fiancée there—" at least two score feminine faces fell with an almost audible thud—"to establish our home, but I hope I'll be pushing on ahead with a calibration cruiser. I have a theory that our Captain's drive may involve some navigational difficulties. And I'll be riding my hobby the while; the arrangement suits me nicely."

Arpe was sure his ears could be seen to be flapping. He was virtually certain that there was no such discipline as microastronomy, and he was perfectly certain that any collimation-cruising (Hammersmith even had the wrong word) the Arpe drive required was going to be done by one Gordon Arpe, except over his dead body.

"*This* man," Celia Gospardi went on implacably, "I'm going to hold, if I have to chase him all over the galaxy. I'll teach him to run away from *me* without making it legal first."

Her fork stabbed a heart of lettuce out of the Lazy Spider and turned it in the gout of Russian dressing the Spider had shot into the air after it. "What does he think he got himself into, anyhow—the Foreign Legion?" she asked nobody in particular. *"Him?* He couldn't find his way out of a super market without a map."

Arpe was gasping like a fish. The girl was smiling warmly at him, from the midst of a cloud of musky perfume against which the ship's ventilators labored in vain. He had never felt less like the captain of a great ship. In another second he would be squirming. He was already blushing.

"Sir . . ."

It was Oestreicher, bending at his ear. Arpe almost broke his tether with gratitude. "Yes, Mr. Oestreicher?"

"We're ready to start dogging down; SV-One has asked us to clear the area a little early, in view of the heavy traffic involved. If you could excuse yourself, we're needed on the bridge."

"Very good. Ladies and gentlemen, please excuse me; I have duties. I hope you'll see the dinner through, and have a good time."

"Is something wrong?" Celia Gospardi said, looking directly into his eyes. His heart went *boomp!* like a form-stamper.

"Nothing wrong," Oestreicher said smoothly from behind him. "There's always work to do in officer's country. Ready, Captain?"

Arpe kicked himself away from the table into the air, avoiding a floating steward only by a few inches. Oestreicher caught up with him in time to prevent his running head-on into the side of a bulkhead.

"We've allowed two hours for the passengers to finish eating and bed down," Oestreicher reported in the control room. "Then we'll start building the field. You're sure we don't need any preparations against acceleration?"

Arpe was recovering; now that the questions were technical, he knew where he was. "No, none at all. The field doesn't mean a thing while it's building. It has to reach a threshold before it takes effect. Once it crosses that point on the curve, it takes effect totally, all at once. Nobody should feel a thing."

"Good. Then we can hit the hammocks for a few hours. I suggest, sir, that Mr. Stauffer take the first watch; I'll take the second; that will leave you on deck when the drive actually fires, if it can be delayed that long. I already have us on a slight retrocurve from SV-One."

"It can be delayed as long as we like. It won't cross the threshold till we close that key."

"That was my understanding," Oestreicher said. "Very

good, sir. Then let's stand the usual watches and get under
way at the fixed time. By then we'll be at apogee so far as
the satellite station is concerned. It would be best to observe
normal routine, right up to the moment when the voyage
itself becomes unavoidably abnormal."

This was wisdom, of course. Arpe could do nothing but
nod, though he doubted very much that he would manage to
get to sleep before his trick came up. The bridge emptied,
except for Stauffer and a j.g. from the Nernst gang, and the
ship quieted.

In the morning, while the passengers were still asleep, Arpe
closed the key.

The *Flyaway II* vanished without a sound.

## 2

Mommy    Mommy    Mommy    Mommy    Mommy
Mommy

I dream I see him  Johnny I love you  he's going down
the ladder into the pit and I can't follow and he's gone already and it's time for the next act

Spaceship  I'm flying it and Bobby can see me and all the
people

Some kind of emergency  but then why not the alarms
Got to ring Stauffer

Daddy?  Daddy?  Bye-bye-see you?  Daddy

Where's the bottle  I knew I shouldn't of gotten sucked
into that game

The wind   always the wind

Falling  falling  why can't I stop falling  will I die if I
stop

Two point eight three four  Two point eight three four  I
keep thinking two point eight three four  that's what the
meter says two point eight three four

Somebody stop that wind  I tell you it talks  I tell you I
hear it  words in the wind

Johnny don't go. I'm riding an elephant and he's trying
to go down the ladder after you and it's going to break

No alarms. All well. But can't think. Can't Mommy ladder
spaceship think for bye-bye-see-you two windy Daddy bottle
seconds straight. What's the bottle trouble game matter anyhow? Where's that two point eight three four physicist, what's-
his-bye-bye-name, Daddy, Johnny, Arpe!

will I die if I stop

I love you

the wind
two point
Mommy
STOP.

STOP. STOP. Arpe. Arpe. Where are you? Everyone else, stop thinking. STOP. We're reading one another's minds. Everyone try to stop before we go nuts. Captain Arpe, do you hear me? Come to the bridge. Arpe, do you hear me?

I hear you. I'm on my way. My God.

You there at the field tension meter

two point eight three four

Yes, you. Concentrate, try not to pay attention to anything else.

Yes, sir. 2.834. 2.834. 2.834.

You people with children, try to soothe them, bed them down again. Mr. Hammersmith!

The wind . . . Yes?

Wake up. We need your help. Oestreicher here. Star deck on the double please. A hey-rube.

But . . . Right, Mr. Oestreicher. On the way.

As the first officer's powerful personality took hold, the raging storm of emotion and dream subsided gradually to a sort of sullen background sea of fear, marked with fleeting whitecaps of hysteria, and Arpe found himself able to think his own thoughts again. There was no doubt about it: everyone on board the *Flyaway II* had become suddenly and totally telepathic.

But what could be the cause? It couldn't be the field. Not only was there nothing in the theory to account for it, but the field had already been effective for nearly an hour, at this same intensity, without producing any such pandemonium.

"My conclusion also," Oestreicher said as Arpe came onto the bridge. "Also you'll notice that we can now see out of the ship, and that the outside sensing instruments are registering again. Neither of those things was true up to a few minutes ago; we went blind as soon as the threshold was crossed."

"Then what's the alternative?" Arpe said. He found that it helped to speak aloud; it diverted him from the undercurrent of the intimate thoughts of everyone else. "It must be characteristic of the space we're in, then, wherever that is. Any clues?"

"There's a sun outside," Stauffer said, "and it has planets.

I'll have the figures for you in a minute. This I can say right away, though: It isn't Alpha Centauri. Too dim."

Somehow, Arpe hadn't expected it to be. Alpha Centauri was in normal space, and this was obviously anything but normal. He caught the figures as they surfaced in Stauffer's mind: Diameter of primary—about a thousand miles (could that possibly be right? Yes, it was correct. But incredible). Number of planets—six. Diameter of outermost planet—about a thousand miles; distance from primary—about 50 million miles.

"What kind of a screwy system is this?" Stauffer protested. "Six planets inside six astronomical units, and the outermost one as big as its sun? It's dynamically impossible."

It certainly was, and yet it was naggingly familiar. Gradually the truth began to dawn on him; there was only one kind of system in which both primary and planet were consistently 1/50,000 of the distance of the outermost orbit. He suppressed it temporarily, partly to see whether or not it was possible to conceal a thought from the others under these circumstances.

"Check the orbital distances, Mr. Stauffer. There should be only two figures involved."

"Two, sir? For six planets?"

"Yes. You'll find two of the bodies occupying the same distance, and the other four at the fifty-million-mile distance."

"Great Scott," Oestreicher said. "Don't tell me we've gotten ourselves inside an atom, sir!"

"Looks like it. Tell me, Mr. Oestreicher, did you get that from my mind, or derive it from what I said?"

"I doped it out," Oestreicher said, puzzled.

"Good; now we know something else: It *is* possible to suppress a thought in this medium. I've been holding the thought 'carbon atom' just below the level of my active consciousness for several minutes."

Oestreicher frowned, and thought: *That's good to know, it increases the possibility of controlling panic and . . .* Slowly, like a sinking ship, the rest of the thought went under. The first officer was practicing.

"You're right about the planets, sir," Stauffer reported. "I suppose this means that they'll all turn out to be the same size, and that there'll be no ecliptic, either."

"Necessarily. They're electrons. That 'sun' is the nucleus."

"But how did it happen?" Oestreicher demanded.

"I can only guess. The field gives us negative mass. We've

never encountered negative mass in nature anywhere but in the microcosm. Evidently that's the only realm where it *can* exist—ergo, as soon as we attained negative mass, we were collapsed into the microcosm."

"Great," Oestreicher grunted. "Can we get out, sir?"

"I don't know. Positive mass is allowable in the microcosm, so if we turned off the field, we might just keep right on staying here. We'll have to study it out. What interests me more right now is this telepathy; there must be some rationale for it."

He thought about it. Until now, he had never believed in telepathy at all; its reported behavior in the macrocosm had been so contrary to all known physical laws that it had been easier to assume that it didn't exist. But the laws of the macrocosm didn't apply down here; this was the domain of quantum mechanics—though telepathy didn't obey that scholium either. Was it possible that the "parapsychological" fields were a part of the fine structure of this universe, as the electromagnetic fields of this universe itself were the fine structure of the macrocosm? If so, any telepathic effects that turned up in the macrocosm would be traces only, a leakage or residuum, fleeting and wayward, beyond all hope of control. . . .

Oestreicher, he noticed, was following his reasoning with considerable interest. "I'm not used to thinking of electrons as having any fine structure," he said.

"Well, all the atomic particles have spin, and to measure that, you have to have some kind of point *on* the particle being translated from one position in space to another—at least by analogy. I would say that the analogy's established now; all we have to do is look out the port."

"You mean we might land on one of those things, sir?" Stauffer asked.

"I should think so," Arpe said, "if we think there's something to be gained by it. I'll leave that up to Mr. Oestreicher."

"Why not?" Oestreicher said, adding, to Arpe's surprise, "The research chance alone oughtn't to be passed up."

Suddenly, the background of fear, which Arpe had more and more become able to ignore, began to swell ominously; huge combers of pure panic were beginning to race over it.

"Oof," Oestreicher said. "We weren't covering enough—we forgot that they could pick up every unguarded word we said. And they don't like the idea."

They didn't. Individual thoughts were hard to catch, but

the main tenor was plain. These people had signed up to go to Centaurus, and that was where they wanted to go. The good possibility that they were trapped on the atomic-size level was terrifying enough, but taking the further risk of landing on an electron . . .

Abruptly Arpe felt, almost without any words to go with it, the raw strength of Hammersmith throwing itself Canute-like against the tide. The explorer's mind had not been in evidence at all since the first shock; evidently he had quickly discovered for himself the trick of masking. For a moment the sheer militancy of Hammersmith's counterstroke seemed to have a calming effect. . . .

One thread of pure terror lifted above the mass. It was Celia Gospardi; she had just awakened, and her shell of bravado had been stripped completely. Following that sound-less scream, the combers of panic became higher, more rapid. . . .

"We'll have to do something about that woman," Oestreicher said tensely. Arpe noted with interest that he was masking the thought he was speaking, quite a difficult technical trick; he tried to mask it also in the reception. "She's going to throw the whole ship into an uproar. You were talking to her at some length last night, sir; maybe you'd better try."

"All right," Arpe said reluctantly, taking a step toward the door. "I gather she's still in her——"

*Flup!*

Celia Gospardi *was* in her stateroom.

So was Captain Arpe.

She stifled a small vocal scream as she recognized him. "Don't be alarmed," he said quickly, though he was almost as alarmed as she was. "Listen, Mr. Oestreicher and everybody else: be careful about making any sudden movements with some definite destination in mind. You're likely to arrive there without having crossed the intervening distance. It's a characteristic of the space we're in."

*I read you, sir. So teleportation is an energy-level jump? That could be nasty, all right.*

"It's—nice of you—to try to—quiet me," the girl said timidly. Arpe noticed covertly that she could not mask worth a damn. He would have to be careful in what he said, for she would effectively make every word known throughout the ship. It was too bad, in a way. Attractive as she was in her public role, she was downright beautiful when frightened.

"Please do try to keep a hold on yourself, Miss Gospardi," he said. "There really doesn't seem to be any immediate danger. The ship is sound and her mechanisms are all operating as they should. We have supplies for a full year, and unlimited power; we ought to be able to get away. There's nothing to be frightened about."

"I can't help it," she said desperately. "I can't even think straight. My thoughts keep getting all mixed up with everybody else's."

"We're all having that trouble to some extent," Arpe said. "If you concentrate, you'll find that you can filter the other thoughts out about ninety per cent. And you'll have to try, because if you remain frightened you'll panic other people—especially the children. They're defenseless against adult emotions even *without* telepathy."

"I—I'll try."

"Good for you." With a slight smile, he added, "After all, if you think as little of your fifth husband as you say, you should welcome a little delay en route."

It was entirely the wrong thing to say. At once, way down at the bottom of her mind, a voice cried out in soundless anguish: *But I love him!*

Tears were running down her cheeks. Helplessly, Arpe left.

He walked carefully, in no hurry to repeat the unnerving teleportation jump. In the main companionway he was waylaid by a junior officer almost at once.

"Excuse me, sir. I have a report here from the ship's surgeon. Dr. Hoyle said it might be urgent and that I'd better bring it to you personally."

"Oh. All right, what is it?"

"Dr. Hoyle's compliments, sir, and he suggests that oxygen tension be checked. He has an acute surgical emergency—a passenger—which suggests that we may be running close to nine thousand."

Arpe tried to think about this, but it did not convey very much to him, and what it did convey was confusing. He knew that space ships, following a tradition laid down long ago in atmospheric flight, customarily expressed oxygen tension in terms of feet of altitude on Earth; but 9000 feet—though it would doubtless cause some discomfort—did not seem to represent a dangerously low concentration. And he could see no connection at all between a slightly depleted oxygen level and an acute surgical emergency. Besides, he was too flustered over Celia Gospardi.

The interview had not ended at all the way he had hoped. But perhaps it was better to have left her grief-stricken than panic-stricken. Of course, if she broadcast her grief all over the ship, there were plenty of other people to receive it, people who had causes for grief as real as hers.

"Grief inactivates," Oestreicher said as Arpe re-entered the bridge. "Even at its worst, it doesn't create riots. Cheer up, sir. I couldn't have done any better, I'm sure of that."

"Thank you, Mr. Oestreicher," Arpe said, flushing. Evidently he had forgotten to mask; "thinking out loud" was more than a cliché down here. To cover, he proffered Hoyle's confusing message.

"Oh?" Oestreicher strode to the mixing board and scanned the big Bourdon gages with a single sweeping glance. "He's right. We're pushing eight-seven hundred right now. Once we cross ten thousand we'll have to order everybody into masks. I *thought* I was feeling a little light-headed. Mr. Stauffer, order an increase in pressure, and get the bubble crew going, on double."

"Right." Stauffer shot out.

"Mr. Oestreicher, what's this all about?"

"We've sprung a major leak, sir—or, more likely, quite a few major leaks. We've got to find out where all this air is going. We may have killed Hoyle's patient already."

Arpe groaned. Surprisingly, Oestreicher grinned.

"Everything leaks," he said in a conversational tone. "That's the first law of space. On the Mars run, when we disliked a captain, we used to wish him an interesting trip. This one is interesting."

"You're a psychologist, Mr. Oestreicher," Arpe said, but he managed to grin back. "Very well; what's the program now? I feel some weight."

"We were making a rocket approach to the nearest electron, sir, and we seem to be moving. I see no reason why we should suspend that. Evidently the Third Law of Motion isn't invalid down here."

"Which is a break," Stauffer said gloomily from the door. "I've got the bubble crew moving, Mr. Oestreicher, but it'll take a while. Captain, what are we seeing by? Gamma waves? Space itself doesn't seem to be dark here."

"Gamma waves are too long," Arpe said. "Probably de Broglie waves. The illuminated sky is probably a demonstration of Obler's Paradox: it's how *our* space would look if the stars were evenly scattered throughout. That makes me think we must be inside a fairly large body of matter. And the nearest one was SV-One."

"Oh-ho," Stauffer said. "And what happens to us when a cosmic ray primary comes charging through here and disrupts our atom?"

Arpe smiled. "You've got the answer to that already. Have you detected any motion in this electron we're approaching?"

"Not much—just normal planetary motion. About fourteen miles a second—expectable for the orbit."

"Which wouldn't be expectable at all unless we were living on an enormously accelerated time scale. By our home time scale we haven't been here a billionth of a second yet. We could spend the rest of our lives here without seeing a free neutron or a cosmic primary."

"That's a relief," Stauffer said; but he sounded a little dubious.

They fell silent as the little world grew gradually in the ports. There was no visible surface detail on it, and the albedo was high. As they came closer, the reasons for both effects became evident, for with each passing moment the outlines of the body grew fuzzier. It seemed to be imbedded in a sort of thick haze.

"Close enough," Oestreicher ruled. "We can't land the *Flyaway* anyhow; we'll have to put a couple of people off in a tender. Any suggestions, sir?"

"I'm going," Arpe said immediately. "I wouldn't miss an opportunity like this for anything."

"Can't blame you, sir," Oestreicher said. "But that body doesn't look like it has any solid core. What if you just sank right through to the center?"

"That's not likely," Arpe said. "I've got a small increment of negative mass, and I'll retain it by picking up the ship's field with an antenna. The electron's light, but what mass it has is positive; in other words, it will repel me slightly. I won't sink far."

"Well then, who's to go with you?" Oestreicher said, masking every word with great care. "One trained observer should be enough, but you'll need an anchor man. I'm astonished that we haven't heard from Hammersmith already—have you noticed how tightly he shut down as soon as this subject came up?"

"So he did," Arpe said, baffled. "I haven't heard a peep out of him for the last hour. Well, that's his problem; maybe he had enough after Titan."

"How about Miss Gospardi?" Stauffer suggested. "It seems to reassure her to be with you, Captain, and it'll give her something new to think about. And it'll take an incipient panic center out of the ship long enough to let the other people calm down."

"Good enough," Arpe said. "Mr. Stauffer, order the gig broken out."

## 3

The little world had a solid surface, after all, though it blended so gradually into the glittering haze of its atmosphere that it was very hard to see. Arpe and the girl seemed to be walking waist-deep in some swirling, opalescent substance that was bearing a colloidal metallic dust, like minute sequins. The faint repulsions against their space suits could not be felt as such; it seemed instead that they were walking in a gravitational field about a tenth that of the Earth.

"It's terribly quiet," Celia said.

The suit radios, Arpe noted, were not working. Luckily, the thought-carrying properties of the medium around them were unchanged.

"I'm not at all sure that this stuff would carry sound," he answered. "It isn't a gas as we know it, anyhow. It's simply a manifestation of indefiniteness. The electron never knows exactly where it is; it just trails off at its boundaries into not being anywhere in particular."

"Well, it's eerie. How long do we have to stay here?"

"Not long. I just want to get some idea of what it's like."

He bent over. The surface, he saw, was covered with fine detail, though again he was unable to make much sense of it. Here and there he saw tiny, crooked rills of some brilliantly shiny substance, rather like mercury, and—yes, there was an irregular puddle of it, and it showed a definite meniscus. When he pushed his finger into it, the puddle dented deeply, but it did not break and wet his glove. Its surface tension must be enormous; he wondered if it were made entirely of identical subfundamental particles. The whole globe seemed to be covered by a network of these shiny threads.

Now that his eyes were becoming acclimated, he saw that the "air," too, was full of these shining veins, making it look distinctly marbled. The veins offered no impediment to their walking; somehow, there never seemed to be any in their immediate vicinity, though there were always many of them just ahead. As the two moved, their progress seemed to be accompanied by vagrant, small emotional currents, without visible cause or source, too fugitive to identify.

"What is that silvery stuff?" Celia demanded fearfully.

"Celia, I haven't the faintest idea. What kind of particle could possibly be submicroscopic to an electron? It'd take a century of research right here on the spot to work up even an educated guess. This is all strange and new, utterly outside any experience man has ever had. I doubt that any words exist to describe it accurately."

The ground, too, seemed to vary in color. In the weak light it was hard to tell what the colors were. The variations appeared as shades of gray, with a bluish or greenish tinge here and there.

The emotional waves became a little stronger, and suddenly Arpe recognized the dominant one.

It was pain.

On a hunch, he turned suddenly and looked behind him. A twin set of broad black bootprints, as solid and sharply defined as if they had been painted, were marked out on the colored patches.

"I don't like the look of that," he said. "Our ship itself is almost of planetary mass in this system, and we're far too big for this planet. How do we know what all this fine detail means? But we're destroying it wherever we step, all the same. Forests, cities, the cells of some organism, something unguessable—we've got to go back right now."

"Believe me, I'm willing," the girl said.

The oldest footprints, those that they had made getting out of the tender, were beginning to grow silvery at the edges, as though with hoarfrost, or with whatever fungus might attack a shadow. Or was it seepage of the same substance that made up the rills? Conjecture multiplied endlessly without answer here. Arpe hated to think of the long oval blot the tender itself would leave behind on the landscape. He could only hope that the damage would be self-repairing; there was something about this place that was peculiarly . . . organic.

He lifted the tender quickly and took it out of the opalescent atmosphere with a minimum of ceremony, casting ahead for guidance to pick up the multifarious murmur of the minds on board the *Flyaway II*.

Only when he noticed that he was searching the sky visually

for the ship did he realize that he was not getting anything.

"Celia? You can hear me all right telepathically, can't you?"

"Clear as a bell. It makes me feel much better, Captain."

"Then what's wrong with the ship? I don't pick up a soul."

She frowned. "Why, neither do I. Where . . ."

Arpe pointed ahead. "There she is, right where we left her. We could hear them all well enough at this distance when we were on the way down. Why can't we now?"

He gunned the tender, all caution forgotten. His arrival in the *Flyaway II*'s air lock was noisy, and he lost several minutes jockeying the little boat into proper seal. They both fell out of it in an inelegant scramble.

There was nobody on board the *Flyaway II*. Nobody but themselves.

The telepathic silence left no doubt in Arpe's or Celia's mind, but they searched the huge vessel thoroughly to make sure. It was deserted.

"Captain!" Celia cried. Her panic was coming back full force. "What happened? Where could they have gone? There isn't any place——"

"I know there isn't. I don't know. Calm down a minute, Celia, and let me think." He sat down on a stanchion and stared blindly at the hull for a moment. Breathing the thinning air was a labor in itself; he found himself wishing they had not shucked their suits. Finally he got up and went back to the bridge, with the girl clinging desperately to his elbow.

Everything was in order. It was as if the whole ship had been deserted simultaneously in an instant. Oestreicher's pipe sat snugly in its clip by the chart board; though it was empty of any trace of the self-oxygenating mixture Oestreicher's juniors had dubbed "Old Gunpowder," the bowl was still hot.

"It can't have happened more than half an hour ago," he whispered. "As if they all did a jump at once—like the one that put me into your stateroom. But where to?"

Suddenly it dawned on him. There was only one answer. Of course they had gone nowhere.

"What is it?" Celia cried. "I can see what you're thinking, but it doesn't make sense!"

"It makes perfect sense—in *this* universe," he said grimly. "Celia, we're going to have to work fast, before Oestreicher makes some stab in the dark that might be irrevocable.

Luckily everything's running as though the crew were still here to tend it—which in fact happens to be true—so maybe two of us will be enough to do what we have to do. But you're going to have to follow instructions fast, accurately, and without stopping for an instant to ask questions."

"What are you going to do?"

"Shut down the field. No, don't protest, you haven't the faintest idea what that means, so you've no grounds for protest. Sit down at that board over there and watch my mind every instant. The moment I think of what you're to do next, do it. Understand?"

"No, but——"

"You understand well enough. All right, let's go."

Rapidly he began to step down the Nernst current going into the field generators, mentally directing Celia in the delicate job of holding the fusion sphere steady against the diminished drain. Within a minute he had the field down to just above the threshold level; the servos functioned without a hitch, and so, not very much to his surprise, did those aspects of the task which were supposed to be manned at all times.

"All right, now I'm going to cut it entirely. There'll be a big backlash on your board. See that master meter right in front of you at the head of the board? The black knob marked 'Back EMF' is cued to it. When I pull this switch, the meter will kick over to some reading above the red line. At the same instant, you roll the knob down to *exactly* the same calibration. If you back it down too far, the Nernst will die and we'll have no power at all. If you don't go down far enough, the Nernst will detonate. You've got to catch it on the nose. Understand?"

"I—think so."

"Good," he said. He hoped it would be good. Normally the rolloff was handled wholly automatically, but by expending the energy evenly into the dying field; they did not dare to chance that here. He could only pray that Celia's first try would be fast. "Here we go. Five seconds, four, three, two, one, *cut.*"

Celia twisted the dial.

For an instant, nothing happened. Then——

Pandemonium.

"Nernst crew chief, report! What are you doing? No orders were——"

"Captain! Miss Gospardi! Where did you spring from?"

This was Oestreicher. He was standing right at Arpe's elbow.

"Stars! Stars!" Stauffer was shouting simultaneously. "Hey, look! Stars! We're *back!*"

There was a confused noise of many people shouting in the belly of the *Flyaway II*. But in Arpe's brain there was blessed silence; the red foaming of raw thoughts by the hundreds was no more. His mind was his own again.

"Good for you, Celia," he said. It was a sort of prayer. "We were in time."

"How did you do it, sir?" Oestreicher was saying. "We couldn't figure it out. We were following your exploration of the electron from here, and suddenly the whole planet just vanished. So did the whole system. We were floating in another atom entirely. We thought we'd lost you for good."

Arpe grinned weakly. "Did you know that you'd left the ship behind when you jumped?"

"But—impossible, sir. It was right here all the time."

"Yes, that too. It was exercising its privilege to be in two places at the same time. As a body with negative mass, it had some of the properties of a Dirac hole; as such, it had to be echoed somewhere else in the universe by an electron, like a sink and a source in calculus. Did you wind up in one of the shells of the second atom?"

"We did," Stauffer said. "We couldn't move out of it, either."

"That's why I killed the field," Arpe explained. "I couldn't know what you would do under the circumstances, but I *was* pretty sure that the ship would resume its normal mass when the field went down. A mass that size, of course, can't exist in the microcosm, so the ship had to snap back. And in the macrocosm it isn't possible for a body to be in two places at the same time. So here we are, gentlemen—reunited."

"Very good, sir," Stauffer said; but the second officer's voice seemed to be a little deficient in hero worship. "But where is here?"

"Eh? Excuse me, Mr. Stauffer, but don't you know?"

"No, sir," Stauffer said. "All I can tell you is that we're nowhere near home, and nowhere near the Centauri stars, either. We appear to be lost, sir."

His glance flicked over to the Bourdon gages.

"Also," he added quietly, "we're still losing air."

The general alarm had alarmed nobody but the crew, who alone knew how rarely it was sounded. As for the bubble gang, the passengers who knew what that meant mercifully kept their mouths shut—perhaps Hammersmith had blustered them into silence—and the rest, reassured at seeing the stars again, were only amused to watch full-grown, grim-looking men stalking the corridors blowing soap bubbles into the air. After a while, the bubble gang vanished; they were working between the hulls.

Arpe was baffled and restive. "Look here," he said suddenly. "This surgical emergency of Hoyle's—I'd forgotten about it, but it seems to have some bearing on this air situation. Let's——"

"He's on his way, sir," Oestreicher said. "I put a call on the bells for him as soon as—ah, here he is now."

Hoyle was a plump, smooth-faced man with a pursed mouth and an expression of perpetual reproof. He looked absurd in his naval whites. He was also four times a Haber medal winner for advances in space medicine.

"It was a ruptured spleen," he said primly. "A dead give-away that we were losing oxygen. I was operating when I had the captain called, or I'd have been more explicit."

"Aha," Oestreicher said. "Your patient's a Negro, then."

"A female Negro—an eighteen-year-old girl, and incidentally one of the most beautiful women I've seen in many, many years."

"What has her color got to do with it?" Arpe demanded, feeling somewhat petulant at Oestreicher's obvious instant comprehension of the situation.

"Everything," Hoyle said. "Like many people of African extraction, she has sicklemia—a hereditary condition in which some of the red blood cells take on a characteristic sicklelike shape. In Africa it was pro-survival, because sicklemic people are not so susceptible to malaria as are people with normal erythrocytes. But it makes them less able to take air that's poor in oxygen—that was discovered back in the 1940s, during the era of unpressurized high altitude airplane flight. It's nothing that can't be dealt with by keeping sufficient oxygen in the ambient air, but . . ."

"How is she?" Arpe said.

"Dying," Hoyle said bluntly. "What else? I've got her in a tent but we can't keep that up forever. I need normal pressure in my recovery room—or if we can't do that, get her back to Earth *fast*."

He saluted sloppily and left. Arpe looked helplessly at Stauffer, who was taking spectra as fast as he could get them onto film, which was far from fast enough for Arpe, let alone the computer. The first attempt at orientation—Schmidt spherical films of the apparent sky, in the hope of identifying at least one constellation, however distorted—had come to nothing. Neither the computer nor any of the officers had been able to find a single meaningful relationship.

"Is it going to do us any good if we do find the Sun?" Oestreicher said. "If we make another jump, aren't we going to face the same situation?"

"Here's S Doradus," Stauffer announced. "That's a beginning, anyhow. But it sure as hell isn't in any position I can recognize."

"We're hoping to find the source of the leak," Arpe reminded the first officer. "But if we don't, I think I can calculate a fast jump—in-again-out-again. I hope we won't have to do it, though. It would involve shooting for a very heavy atom—heavy enough to be unstable——"

"Looking for the Sun?" a booming, unpleasantly familiar voice broke in from the bulkhead. It was Hammersmith, of course. Dogging his footsteps was Dr. Hoyle, looking even more disapproving than ever.

"See here, Mr. Hammersmith," Arpe said. "This is an emergency. You've got no business being on the bridge at all."

"You don't seem to be getting very far with the job," Hammersmith observed, with a disparaging glance at Stauffer. "And it's my life as much as it's anybody else's. It's high time I gave you a hand."

"We'll get along," Oestreicher said, his face red. "Your stake in the matter is no greater than any other passenger's——"

"Ah, that's not quite true," Dr. Hoyle said, almost regretfully. "The emergency is medically about half Mr. Hammersmith's."

"Nonsense," Arpe said sharply. "If there's any urgency beyond what affects us all, it affects your patient primarily."

"Yes, quite so," Dr. Hoyle said, spreading his hands. "She is Mr. Hammersmith's fiancée."

After a moment, Arpe discovered that he was angry—not with Hammersmith, but with himself, for being stunned by the announcement. There was nothing in the least unlikely about such an engagement, and yet it had never entered his head even as a possibility. Evidently his unconscious still had prejudices he had extirpated from his conscious mind thirty-five years ago.

"Why have you been keeping it a secret?" he asked slowly.

"For Helen's protection," Hammersmith said, with considerable bitterness. "On Centaurus we may get a chance at a reasonable degree of privacy and acceptance. But if I'd kept her with me on the ship, she'd have been stared at and whispered over for the entire trip. She preferred to stay below."

An ensign came in, wearing a space suit minus the helmet, and saluted clumsily. After he got the space suit arm up, he just left it there, resting his arm inside it. He looked like a small doll some child had managed to stuff inside a larger one.

"Bubble team reporting, sir," he said. "We were unable to find any leaks, sir."

"You're out of your mind," Oestreicher said sharply. "The pressure is still dropping. There's a hole somewhere you could put your head through."

"No, sir," the ensign said wearily. "There are no such holes. The entire ship is leaking. The air is going right out through the metal. The rate of loss is perfectly even, no matter where you test it."

"Osmosis!" Arpe exclaimed.

"What do you mean, sir?" Oestreicher said.

"I'm not sure, Mr. Oestreicher. But I've been wondering all along—I guess we all have—just how this whole business would affect the ship structurally. Evidently it weakened the molecular bonds of everything on board—and now we have good structural titanium behaving like a semipermeable membrane! I'll bet it's specific for oxygen, furthermore; a 20 per cent drop in pressure is just about what we're getting here."

"What about the effect on people?" Oestreicher said.

"That's Dr. Hoyle's department," Arpe said. "But I rather doubt that it affects living matter. That's in an opposite state of entropy. But when we get back, I want to have the ship measured. I'll bet it's several meters bigger in both length and girth than it was when it was built."

"*If* we get back," Oestreicher said, his brow dark.

"Is this going to put the kibosh on your drive?" Stauffer asked gloomily.

"It's going to make interstellar flight pretty expensive,"

Arpe admitted. "It looks like we'll have to junk a ship after one round trip."

"Well, we effectively junked the *Flyaway I* after one *one-way* trip," Oestreicher said reflectively. "That's progress, of a sort."

"Look here, all this jabber isn't getting us anywhere," Hammersmith said. "Do you want me to bail you out, or not? If not, I'd rather be with Helen than standing around listening to you."

"What do you propose to do," Arpe said, finding it impossible not to be frosty, "that we aren't doing already?"

"Teach you your business," Hammersmith said. "I presume you've established our distance from S Doradus for a starter. Once I have that, I can use the star as a beacon, to collimate my next measurements. Then I want the use of an image amplifier, with a direct-reading microvoltmeter tied into the circuit; you ought to have such a thing, as a routine instrument."

Stauffer pointed it out silently.

"Good." Hammersmith sat down and began to scan the stars with the amplifier. The meter silently reported the light output of each, as minute pulses of electricity. Hammersmith watched it with a furious intensity. At last he took off his wrist chronometer and begun to time the movements of the needle with the stop watch.

"Bull's-eye," he said suddenly.

"The Sun?" Arpe asked, unable to keep his tone from dripping with disbelief.

"No. That one is DQ Herculis—an old nova. It's a microvariable. It varies by four hundredths of a magnitude every sixty-four seconds. Now we have two stars to fill our parameters; maybe the computer could give us the Sun from those? Let's try it, anyhow."

Stauffer tried it. The computer had decided to be obtuse today. It did, however, narrow the region of search to a small sector of sky, containing approximately sixty stars.

"Does the Sun do something like that?" Oestreicher said. "I knew it was a variable star in the radio frequencies, but what about visible light?"

"If we could mount an RF antenna big enough, we'd have the Sun in a moment," Hammersmith said in a preoccupied voice. "But with light it's more complicated. . . . Um. If *that's* the Sun, we must be even farther away from it than I thought.

Dr. Hoyle, will you take my watch, please, and take my pulse?"

"Your pulse?" Hoyle said, startled. "Are you feeling ill? The air is——"

"I feel fine, I've breathed thinner air than this and lived," Hammersmith said irritably. "Just take my pulse for a starter, then take everyone else's here and give me the average. I'd use the whole shipload if I had the time, but I don't. If none of you experts knows what I'm doing I'm not going to waste what time I've got explaining it to you now. Goddam it, there are lives involved, remember?"

His lips thinned, Arpe nodded silently to Hoyle; he did not trust himself to speak. The physician shrugged his shoulders and began collecting pulse rates, starting with the big explorer. After a while he had an average and passed it to Hammersmith on a slip of paper torn from his report book.

"Good," Hammersmith said. "Mr. Stauffer, please feed this into Bessie there. Allow for a permitted range of variation of two per cent, and bleed the figure out into a hundred and six increments and decrements each; then tell me what the percentage is now. Can do?"

"Simple enough." Stauffer programmed the tape. The computer jammered out the answer almost before the second officer had stopped typing; Stauffer handed the strip of paper over to Hammersmith.

Arpe watched with reluctant fascination. He had no idea what Hammersmith was doing, but he was beginning to believe that there was such a science as microastronomy after all.

Thereafter, there was a long silence while Hammersmith scanned one star after another. At last he sighed and said:

"There you are. This ninth magnitude job I'm lined up on now. That's the Sun. Incidentally we are a little closer to Alpha Centauri than we are from home—though God knows we're a long way from either."

"How can you be sure?" Arpe said.

"I'm not sure. But I'm as sure as I can be at this distance. Pick the one you want to go to, make the jump, and I'll explain afterwards. We can't afford to kill any more time with lectures."

"No," Arpe said. "I will do no such thing. I'm not going to throw away what will probably be our only chance—the ship isn't likely to stand more than one more jump—on a calculation that I don't even know the rationale of."

"And what's the alternative?" Hammersmith demanded, sneering slightly. "Sit here and die of anoxia—and just sheer damn stubbornness?"

"I am the captain of this vessel," Arpe said, flushing. "We do not move until I get a satisfactory explanation of your pretensions. Do you understand me? That's my order; it's final."

For a few moments the two men glared at each other, stiff-necked as idols, each the god of his own pillbox-universe.

Hammersmith's eyelids drooped. All at once, he seemed too tired to care.

"You're wasting time," he said. "Surely it would be faster to check the spectrum."

"Excuse me, Captain," Stauffer said excitedly. "I just did that. And I think that star *is* the Sun. It's about eight hundred light-years away——"

*"Eight hundred light-years!"*

"Yes, sir, at least that. The spectral lines are about half missing, but all the ones that are definite enough to measure match nicely with the Sun's. I'm not so sure about the star Mr. Hammersmith identifies as A Centaurus, but at the very least it's a spectroscopic double, and it *is* about fifty light-years closer."

"My God," Arpe muttered. "Eight hundred."

Hammersmith looked up again, his expression curiously like that of a St. Bernard whose cask of brandy has been spurned. "Isn't that sufficient?" he said hoarsely. "In God's name, let's get going. She's dying while we stand around here nit-picking!"

"No rationale, no jump," Arpe said stonily. Oestreicher shot him a peculiar glance out of the corners of his eyes. In that moment, Arpe felt his painfully accumulated status with the first officer shatter like a Prince Rupert's drop; but he would not yield.

"Very well," Hammersmith said gently. "It goes like this. The Sun is a variable star. With a few exceptions, the pulses don't exceed the total average emission—the solar constant— by more than two per cent. The over-all period is 273 months. Inside that, there are at least sixty-three subordinate cycles. There's one of 212 days. Another one lasts only a fraction over six and a half days—I forget the exact period, but it's 1/1250 of the main cycle, if you want to work it out on Bessie there."

"I guessed something like that," Arpe said. "But what good does it do us? We have no tables for it——"

"These cycles have effects," Hammersmith said. "The six-and-a-half-day cycle strongly influences the weather on Earth, for instance. And the 212-day cycle is reflected one-for-one *in the human pulse rate.*"

"Oho," Oestreicher said. "Now I see. It's—Captain, this means that we can *never* be lost! Not so long as the Sun is detectable at all, whether we can identify it or not! We're carrying the only beacon we need right in our blood!"

"Yes," Hammersmith said. "That's how it goes. It's better to take an average of all the pulses available, since one man might be too excited to give you an accurate figure. I'm that overwrought myself. I wonder if it's patentable? No, a law of nature, I suppose; besides, too easily infringed, almost like a patent on shaving. . . . But it's true, Mr. Oestreicher. You may go as far afield as you please, but your Sun stays in your blood. You never really leave home."

He lifted his head and looked at Arpe with hooded, blood-shot eyes.

"Now can we go, please?" he said, almost in a whisper. "And, Captain—if this delay has killed Helen, you will answer to me for it, if I have to chase you to the smallest, most remote star that God ever made."

Arpe swallowed. "Mr. Stauffer," he said, "prepare for jump."

"Where to, sir?" the second officer said. "Back home—or to destination?"

And there was the crux. After the next jump the *Fly-away II* would not be spaceworthy any more. If they used it up making Centaurus, they would be marooned; they would have made their one round trip one-way. Besides . . . *your drive is more important than anything else on board. Get the passengers where they want to go by all means if it's feasible, but if it isn't, the government wants that drive back. . . . Understand?*

"We contracted with the passengers to go to Centaurus," Arpe said, sitting down before the computer. "That's where we'll go."

"Very good, sir," Oestreicher said. They were the finest three words Arpe had ever heard in his life.

The Negro girl, exquisite even in her still and terrible coma, was first off the ship into the big ship-to-shore ferry.

Hammersmith went with her, his big face contorted with anguish.

Then the massive job of evacuating everybody else began. Everyone—passengers and ship's complement alike—was wearing a mask now. After the jump through the heavy cosmic-ray primary that Arpe had picked, a stripped nucleus which happened to be going toward Centaurus anyhow, the *Flyaway II* was leaking air as though she were made of something not much better than surgical gauze. She was through.

Oestreicher turned to Arpe and held out his hand. "A great achievement, sir," the first officer said. "It'll be cut and dried into a routine after it's collimated—but they won't even know that back home until the radio word comes through, better than four years from now. I'm glad I was along while it was still new."

"Thank you, Mr. Oestreicher. You won't miss the Mars run?"

"They'll need interplanetary captains here too, sir." He paused. "I'd better go help Mr. Stauffer with the exodus."

"Right. Thank you, Mr. Oestreicher."

Then he was alone. He meant to be last off the ship; after living with Oestreicher and his staff for so long, he had come to see that traditions do not grow from nothing. After a while, however, the bulkhead lock swung heavily open, and Dr. Hoyle came in.

"Skipper, you're bushed. Better knock it off."

"No," Arpe said in a husky voice, not turning away from watching through the viewplate the flaming departure of the ferry for the green and brown planet, so wholly Earthlike except for the strange shapes of its continents, a thousand miles below. "Hoyle, what do you think? Has she still got a chance?"

"I don't know. It will be nip and tuck. Maybe. Wilson—he was ship's surgeon on the *Flyaway I*—will pick her up as she lands. He's not young any more, but he was as good as they came; and with a surgeon it isn't age that matters, it's how frequently you operate. But . . . she was on the way out for a long time. She may be a little . . ."

He stopped.

"Go on," Arpe said. "Give it to me straight. I know I was wrong."

"She was low on oxygen for a long time," Hoyle said, without looking at Arpe. "It may be that she'll be a little simple-minded when she recovers. Or it may not; there's no predicting these things. But one thing's for sure; she'll never dare go into space again. Not even back to Earth. The next slight drop in oxygen tension will kill her. I even advised against airplanes for her, and Wilson concurs."

Arpe swallowed. "Does Hammersmith know that?"

"Yes," Hoyle said, "he knows it. But he'll stick with her. He loves her."

The ferry carrying the explorer and his fiancée, and Captain Willoughby's daughter and her Judy, and many others, was no longer visible. Sick at heart, Arpe watched Centaurus III turn below him.

That planet was the gateway to the stars—for everyone on it but Daryon and Helen Hammersmith. The door that had closed behind them when they had boarded the ferry was for them no gateway to any place. It was only the door to a prison.

But it was also, Arpe realized suddenly, a prison which would hold a great teacher—not of the humanities, but of Humanity. Arpe, not so imprisoned, had no such thing to teach.

It was true that he knew how to do a great thing—how to travel to the stars. It was true that he had taken Celia Gospardi and the others where they had wanted to go. It was true that he was now a small sort of hero to his crew; and it was true that he—Dr. Gordon Arpe, sometime laboratory recluse, sometime *ersatz* space-ship captain, sometime petty hero, had been kissed good-bye by a 3-V star.

But it was also over. From now on he could do no more than sit back and watch others refine the Arpe drive; the four-year communication gap between Centaurus and home would shut him out of those experiments as though he were a Cro-Magnon Man—or Daryon Hammersmith. When next Arpe saw an Earth physicist, he wouldn't have the smallest chance of understanding a word the man said.

That was a prison, too; a prison Capt. Gordon Arpe had fashioned himself, and then had thrown away the key.

"Beg pardon, Captain?"

"Oh. Sorry, Dr. Hoyle. Didn't realize you were still here." Arpe looked down for the last time on the green-and-brown planet, and drew a long breath. "I said, 'So be it.'"

# Beep

JOSEF FABER lowered his newspaper slightly. Finding the girl on the park bench looking his way, he smiled the agonizingly embarrassed smile of the thoroughly married nobody caught bird-watching, and ducked back into the paper again.

He was reasonably certain that he looked the part of a middle-aged, steadily employed, harmless citizen enjoying a Sunday break in the bookkeeping and family routines. He was also quite certain, despite his official instructions, that it wouldn't make the slightest bit of difference if he didn't. These boy-meets-girl assignments always came off. Jo had never tackled a single one that had required him.

As a matter of fact, the newspaper, which he was supposed to be using only as a blind, interested him a good deal more than his job did. He had only barely begun to suspect the obvious ten years ago when the Service had snapped him up; now, after a decade as an agent, he was still fascinated to see how smoothly the really important situations came off. The *dangerous* situations—not boy-meets-girl.

This affair of the Black Horse Nebula, for instance. Some days ago the papers and the commentators had begun to mention reports of disturbances in that area, and Jo's practiced eye had picked up the mention. Something big was cooking.

Today it had boiled over—the Black Horse Nebula had suddenly spewed ships by the hundreds, a massed armada that must have taken more than a century of effort on the part of a whole star cluster, a production drive conducted in the strictest and most fanatical kind of secrecy. . . .

And, of course, the Service had been on the spot in plenty of time. With three times as many ships, disposed with mathematical precision so as to enfilade the entire armada the moment it broke from the nebula. The battle had been a massacre, the attack smashed before the average citizen could even begin to figure out what it had been aimed at—and good had triumphed over evil once more.

Of course.

Furtive scuffings on the gravel drew his attention briefly. He looked at his watch, which said 14:58:03. That was the time, according to his instructions, when boy had to meet girl.

He had been given the strictest kind of orders to let nothing interfere with this meeting—the orders always issued on boy-meets-girl assignments. But, as usual, he had nothing to do but observe. The meeting was coming off on the dot, without any prodding from Jo. They always did.

Of course.

With a sigh, he folded his newspaper, smiling again at the couple—yes, it was the right man, too—and moved away, as if reluctantly. He wondered what would happen were he to pull away the false mustache, pitch the newspaper on the grass, and bound away with a joyous whoop. He suspected that the course of history would not be deflected by even a second of arc, but he was not minded to try the experiment.

The park was pleasant. The twin suns warmed the path and the greenery without any of the blasting heat which they would bring to bear later in the summer. Randolph was altogether the most comfortable planet he had visited in years. A little backward, perhaps, but restful, too.

It was also slightly over a hundred light-years away from Earth. It would be interesting to know how Service headquarters on Earth could have known in advance that boy would meet girl at a certain spot on Randolph, precisely at 14:58:03.

Or how Service headquarters could have ambushed with micrometric precision a major interstellar fleet, with no more preparation than a few days' buildup in the newspapers and video could evidence.

The press was free, on Randolph as everywhere. It reported the news it got. Any emergency concentration of Service ships in the Black Horse area, or anywhere else, would have been noticed and reported on. The Service did not forbid such reports for "security" reasons or for any other reasons. Yet there had been nothing to report but that (a) an armada of staggering size had erupted with no real warning from the Black Horse Nebula, and that (b) the Service had been ready.

By now, it was a commonplace that the Service was always ready. It had not had a defect or a failure in well over two centuries. It had not even had a fiasco, the alarming-sounding technical word by which it referred to the possibility that a boy-meets-girl assignment might not come off.

Jo hailed a hopper. Once inside he stripped himself of the

mustache, the bald spot, the forehead creases—all the make-up which had given him his mask of friendly innocuousness.

The hoppy watched the whole process in the rear-view mirror. Jo glanced up and met his eyes.

"Pardon me, mister, but I figured you didn't care if I saw you. You must be a Service man."

"That's right. Take me to Service HQ, will you?"

"Sure enough." The hoppy gunned his machine. It rose smoothly to the express level. "First time I ever got close to a Service man. Didn't hardly believe it at first when I saw you taking your face off. You sure looked different."

"Have to, sometimes," Jo said, preoccupied.

"I'll bet. No wonder you know all about everything before it breaks. You must have a thousand faces each, your own mother wouldn't know you, eh? Don't you care if I know about your snooping around in disguise?"

Jo grinned. The grin created a tiny pulling sensation across one curve of his cheek, just next to his nose. He stripped away the overlooked bit of tissue and examined it critically.

"Of course not. Disguise is an elementary part of Service work. Anyone could guess that. We don't use it often, as a matter of fact—only on very simple assignments."

"Oh." The hoppy sounded slightly disappointed, as melodrama faded. He drove silently for about a minute. Then, speculatively: "Sometimes I think the Service must have time-travel, the things they pull. . . . Well, here you are. Good luck, mister."

"Thanks."

Jo went directly to Krasna's office. Krasna was a Randolpher. Earth-trained, and answerable to the Earth office, but otherwise pretty much on his own. His heavy, muscular face wore the same expression of serene confidence that was characteristic of Service officials everywhere—even some that, technically speaking, had no faces to wear it.

"Boy meets girl," Jo said briefly. "On the nose and on the spot."

"Good work, Jo. Cigarette?" Krasna pushed the box across his desk.

"Nope, not now. Like to talk to you, if you've got time."

Krasna pushed a button, and a toadstoollike chair rose out of the floor behind Jo. "What's on your mind?"

"Well," Jo said carefully. "I'm wondering why you patted me on the back just now for not doing a job."

"You did a job."

"I did not," Jo said flatly. "Boy would have met girl, whether I'd been here on Randolph or back on Earth. The course of true love always runs smooth. It has in all my boy-meets-girl cases, and it has in the boy-meets-girl cases of every other agent with whom I've compared notes."

"Well, good," Krasna said, smiling. "That's the way we like to have it run. And that's the way we expect it to run. But, Jo, we like to have somebody on the spot, somebody with a reputation for resourcefulness, just in case there's a snag. There almost never is, as you've observed. But—if there were?"

Jo snorted. "If what you're trying to do is to establish preconditions for the future, any interference by a Service agent would throw the eventual result farther *off* the track. I know that much about probability."

"And what makes you think that we're trying to set up the future?"

"It's obvious even to the hoppies on your own planet; the one that brought me here told me he thought the Service had time-travel. It's especially obvious to all the individuals and governments and entire populations that the Service has bailed out of serious messes for centuries, with never a single failure." Jo shrugged. "A man can be asked to safeguard only a small number of boy-meets-girl cases before he realizes, as an agent, that what the Service is safeguarding is the future children of those meetings. Ergo—the Service *knows* what those children are to be like, and has reason to want their future existence guaranteed. What other conclusion is possible?"

Krasna took out a cigarette and lit it deliberately; it was obvious that he was using the maneuver to cloak his response.

"None," he admitted at last. "We have some foreknowledge, of course. We couldn't have made our reputation with espionage alone. But we have obvious other advantages: genetics, for instance, and operations research, the theory of games, the Dirac transmitter—it's quite an arsenal, and of course there's a good deal of prediction involved in all those things."

"I see that," Jo said. He shifted in his chair, formulating all he wanted to say. He changed his mind about the cigarette and helped himself to one. "But these things don't add up to infallibility—and that's a qualitative difference, Kras. Take this affair of the Black Horse armada. The moment the armada appeared, we'll assume, Earth heard about it by Dirac, and started to assemble a counterarmada. But it takes *finite time* to bring together a concentration of ships and men, even if your message system is instantaneous.

"The Service's counterarmada was *already on hand*. It had been building there for so long and with so little fuss that nobody even noticed it concentrating until a day or so before the battle. Then planets in the area began to sit up and take notice, and be uneasy about what was going to break. But not very uneasy; the Service always wins—that's been a statistical fact for centuries. *Centuries*, Kras. Good Lord, it takes almost as long as that, in straight preparation, to pull some of the tricks we've pulled! The Dirac gives us an advantage of ten to twenty-five years in really extreme cases out on the rim of the Galaxy, but no more than that."

He realized that he had been fuming away on the cigarette until the roof of his mouth was scorched, and snubbed it out angrily. "That's a very different thing," he said, "than knowing in a general way how an enemy is likely to behave, or what kind of children the Mendelian laws say a given couple should have. It means that we've some way of reading the future in minute detail. That's in flat contradiction to everything I've been taught about probability, but I have to believe what I see."

Krasna laughed. "That's a very able presentation," he said. He seemed genuinely pleased. "I think you'll remember that you were first impressed into the Service when you began to wonder why the news was always good. Fewer and fewer people wonder about that nowadays; it's become a part of their expected environment." He stood up and ran a hand through his hair. "Now you've carried yourself through the next stage. Congratulations, Jo. You've just been promoted!"

"I have?" Jo said incredulously. "I came in here with the notion that I might get myself fired."

"No. Come around to this side of the desk, Jo, and I'll play you a little history." Krasna unfolded the desktop to expose a small visor screen. Obediently Jo rose and went around the desk to where he could see the blank surface. "I had a standard indoctrination tape sent up to me a week ago, in the expectation that you'd be ready to see it. Watch."

Krasna touched the board. A small dot of light appeared in the center of the screen and went out again. At the same time, there was a small *beep* of sound. Then the tape began to unroll and a picture clarified on the screen.

"As you suspected," Krasna said conversationally, "the Service is infallible. How it got that way is a story that started several centuries back.

## 2

Dana Lje—her father had been a Hollander, her mother born in the Celebes—sat down in the chair which Captain Robin Weinbaum had indicated, crossed her legs, and waited, her blue-black hair shining under the lights.

Weinbaum eyed her quizzically. The conqueror Resident who had given the girl her entirely European name had been paid in kind, for his daughter's beauty had nothing fair and Dutch about it. To the eye of the beholder, Dana Lje seemed a particularly delicate virgin of Bali, despite her Western name, clothing and assurance. The combination had already proven piquant for the millions who watched her television column, and Weinbaum found it no less charming at first hand.

"As one of your most recent victims," he said, "I'm not sure that I'm honored, Miss Lje. A few of my wounds are still bleeding. But I am a good deal puzzled as to why you're visiting me now. Aren't you afraid that I'll bite back?"

"I had no intention of attacking you personally, and I don't think I did," the video columnist said seriously. "It was just pretty plain that our intelligence had slipped badly in the Erskine affair. It was my job to say so. Obviously you were going to get hurt, since you're head of the bureau —but there was no malice in it."

"Cold comfort," Weinbaum said dryly. "But thank you, nevertheless."

The Eurasian girl shrugged. "That isn't what I came here about, anyway. Tell me, Captain Weinbaum—have you ever heard of an outfit calling itself Interstellar Information?"

Weinbaum shook his head. "Sounds like a skip-tracing firm. Not an easy business, these days."

"That's just what I thought when I first saw their letter-head," Dana said. "But the letter under it wasn't one that a private-eye outfit would write. Let me read part of it to you."

Her slim fingers burrowed in her inside jacket pocket and emerged again with a single sheet of paper. It was plain typewriter bond, Weinbaum noted automatically: she had brought only a copy with her, and had left the original of the letter at home. The copy, then, would be incomplete—probably seriously.

"It goes like this: 'Dear Miss Lje: As a syndicated video commentator with a wide audience and heavy responsibilities, you need the best sources of information available. We would like you to test our service, free of charge, in the hope of proving to you that it is superior to any other source of news on Earth. Therefore, we offer below several predictions concerning events to come in the Hercules and the so-called "Three Ghosts" areas. If these predictions are fulfilled 100 per cent—no less—we ask that you take us on as your correspondents for those areas, at rates to be agreed upon later. If the predictions are wrong in *any* respect, you need not consider us further.' "

"H'm," Weinbaum said slowly. "They're confident cusses —and that's an odd juxtaposition. The Three Ghosts make up only a little solar system, while the Hercules area could include the entire star cluster—or maybe even the whole constellation, which is a hell of a lot of sky. This outfit seems to be trying to tell you that it has thousands of field correspondents of its own, maybe as many as the government itself. If so, I'll guarantee that they're bragging."

"That may well be so. But before you make up your mind, let me read you one of the two predictions." The letter rustled in Dana Lje's hand. " 'At 03:16:10, on Year Day, 2090, the Hess-type interstellar liner *Brindisi* will be attacked in the neighborhood of the Three Ghosts system by four——' "

Weinbaum sat bolt upright in his swivel chair. "Let me see that letter!" he said, his voice harsh with repressed alarm.

"In a moment," the girl said, adjusting her skirt composedly. "Evidently I was right in riding my hunch. Let me go on reading: '—by four heavily armed vessels flying the lights of the navy of Hammersmith II. The position of the liner at that time will be at coded co-ordinates 88-A-theta-88-aleph-D and-per-se-and. It will——' "

"Miss Lje," Weinbaum said. "I'm sorry to interrupt you again, but what you've said already would justify me in jailing you at once, no matter how loudly your sponsors might scream. I don't know about this Interstellar Information outfit, or whether or not you did receive any such letter as the one you pretend to be quoting. But I can tell you that you've shown yourself to be in possession of information that only yours truly and four other men are supposed to know. It's already too late to tell you that everything you say may be held against you; all I can say now is, it's high time you clammed up!"

"I thought so," she said, apparently not disturbed in the least. "Then that liner *is* scheduled to hit those co-ordinates, and the coded time co-ordinate corresponds with the predicted Universal Time. Is it also true that the *Brindisi* will be carrying a top-secret communication device?"

"Are you deliberately trying to make me imprison you?" Weinbaum said, gritting his teeth. "Or is this just a stunt, designed to show me that my own bureau is full of leaks?"

"It could turn into that," Dana admitted. "But it hasn't, yet. Robin, I've been as honest with you as I'm able to be. You've had nothing but square deals from me up to now. I wouldn't yellow-screen you, and you know it. If this unknown outfit has this information, it might easily have gotten it from where it hints that it got it: from the field."

"Impossible."

"Why?"

"Because the information in question hasn't even reached my *own* agents in the field yet—it couldn't possibly have leaked as far as Hammersmith II or anywhere else, let alone to the Three Ghosts system! Letters have to be carried on ships, you know that. If I were to send orders by ultrawave to my Three Ghosts agent, he'd have to wait three hundred and twenty-four years to get them. By ship, he can get them in a little over two months. These particular orders have only been under way to him five days. Even if somebody has read them on board the ship that's carrying them, they couldn't possibly be sent on to the Three Ghosts any faster than they're traveling now."

Dana nodded her dark head. "All right. Then what are we left with but a leak in your headquarters here?"

"What, indeed," Weinbaum said grimly. "You'd better tell me who signed this letter of yours."

"The signature is J. Shelby Stevens."

Weinbaum switched on the intercom. "Margaret, look in the business register for an outfit called Interstellar Information and find out who owns it."

Dana Lje said, "Aren't you interested in the rest of the prediction?"

"You bet I am. Does it tell you the name of this communications device?"

"Yes," Dana said.

"What is it?"

"The Dirac communicator."

Weinbaum groaned and turned on the intercom again. "Margaret, send in Dr. Wald. Tell him to drop everything and gallop. Any luck with the other thing?"

"Yes, sir," the intercom said. "It's a one-man outfit, wholly owned by a J. Shelby Stevens, in Rico City. It was first registered this year."

"Arrest him, on suspicion of espionage."

The door swung open and Dr. Wald came in, all six and a half feet of him. He was extremely blond, and looked awkward, gentle, and not very intelligent.

"Thor, this young lady is our press nemesis, Dana Lje. Dana, Dr. Wald is the inventor of the Dirac communicator, about which you have so damnably much information."

"It's out *already?*" Dr. Wald said, scanning the girl with grave deliberation.

"It is, and lots more—*lots* more. Dana, you're a good girl at heart, and for some reason I trust you, stupid though it is to trust anybody in this job. I should detain you until Year Day, videocasts or no videocasts. Instead, I'm just going to ask you to sit on what you've got, and I'm going to explain why."

"Shoot."

"I've already mentioned how slow communication is between star and star. We have to carry all our letters on ships, just as we did locally before the invention of the telegraph. The overdrive lets us beat the speed of light, but not by much of a margin over really long distances. Do you understand that?"

"Certainly," Dana said. She appeared a bit nettled, and Weinbaum decided to give her the full dose at a more rapid pace. After all, she could be assumed to be better informed than the average layman.

"What we've needed for a long time, then," he said, "is some virtually instantaneous method of getting a message from somewhere to anywhere. Any time lag, no matter how small it seems at first, has a way of becoming major as longer and longer distances are involved. Sooner or later we must have this instantaneous method, or we won't be able to get messages from one system to another fast enough to hold our jurisdiction over outlying regions of space."

"Wait a minute," Dana said. "I'd always understood that ultrawave is faster than light."

"Effectively it is; physically it isn't. You don't understand that?"

She shook her dark head.

"In a nutshell," Weinbaum said, "ultrawave is radiation, and all radiation in free space is limited to the speed of light. The way we hype up ultrawave is to use an old application of wave-guide theory, whereby the real transmission of energy is at light speed, but an imaginary thing called "phase velocity" is going faster. But the gain in speed of transmission isn't large—by ultrawave, for instance, we get a message to Alpha Centauri in one year instead of nearly four. Over long distances, that's not nearly enough extra speed."

"Can't it be speeded further?" she said, frowning.

"No. Think of the ultrawave beam between here and Centaurus III as a caterpillar. The caterpillar himself is moving quite slowly, just at the speed of light. But the pulses which pass along his body are going forward faster than he is—and if you've ever watched a caterpillar, you'll know that that's true. But there's a physical limit to the number of pulses you can travel along that caterpillar, and we've already reached that limit. We've taken phase velocity as far as it will go.

"That's why we need something faster. For a long time our relativity theories discouraged hope of anything faster—even the high-phase velocity of a guided wave didn't contradict those theories; it just found a limited, mathematically imaginary loophole in them. But when Thor here began looking into the question of the velocity of propagation of a Dirac pulse, he found the answer. The communicator he developed does seem to act over long distances, *any* distance, instantaneously—and it may wind up knocking relativity into a cocked hat."

The girl's face was a study in stunned realization. "I'm not sure I've taken in all the technical angles," she said. "But if I'd had any notion of the political dynamite in this thing——"

"—you'd have kept out of my office," Weinbaum said grimly. "A good thing you didn't. The *Brindisi* is carrying a model of the Dirac communicator out to the periphery for a final test; the ship is supposed to get in touch with me from out there at a given Earth time, which we've calculated very elaborately to account for the residual Lorentz and Milne transformations involved in overdrive flight, and for a lot of other time phenomena that wouldn't mean anything at all to you.

"If that signal arrives here at the given Earth time, then —aside from the havoc it will create among the theoretical physicists whom we decide to let in on it—we will really have our instant communicator, and can include all of occupied space in the same time zone. And we'll have a terrific advantage over any lawbreaker who has to resort to ultrawave locally and to letters carried by ships over the long haul."

"Not," Dr. Wald said sourly, "if it's already leaked out."

"It remains to be seen how much of it has leaked," Weinbaum said. "The principle is rather esoteric, Thor, and the name of the thing alone wouldn't mean much even to a trained scientist. I gather that Dana's mysterious informant didn't go into technical details . . . or did he?"

"No," Dana said.

"Tell the truth, Dana. I know that you're suppressing some of that letter."

The girl started slightly. "All right—yes, I am. But nothing technical. There's another part of the prediction that lists the number and class of ships you will send to protect the *Brindisi*—the prediction says they'll be sufficient, by the way —and I'm keeping that to myself, to see whether or not it comes true along with the rest. If it does, I think I've hired myself a correspondent."

"If it does," Weinbaum said, "you've hired yourself a jailbird. Let's see how much mind reading J. Whatsit Stevens can do from the subcellar of Fort Yaphank."

### 3

Weinbaum let himself into Stevens's cell, locking the door behind him and passing the keys out to the guard. He sat down heavily on the nearest stool.

Stevens smiled the weak benevolent smile of the very old, and laid his book aside on the bunk. The book, Weinbaum knew—since his office had cleared it—was only a volume of pleasant, harmless lyrics by a New Dynasty poet named Nims.

"Were our predictions correct, Captain?" Stevens said. His voice was high and musical, rather like that of a boy soprano.

Weinbaum nodded. "You still won't tell us how you did it?"

"But I already have," Stevens protested. "Our intelligence

network is the best in the Universe, Captain. It is superior even to your own excellent organization, as events have shown."

"Its results are superior, that I'll grant," Weinbaum said glumly. "If Dana Lje had thrown your letter down her disposal chute, we would have lost the *Brindisi* and our Dirac transmitter both. Incidentally, did your original letter predict accurately the number of ships we would send?"

Stevens nodded pleasantly, his neatly trimmed white beard thrusting forward slightly as he smiled.

"I was afraid so," Weinbaum leaned forward. "Do you have the Dirac transmitter, Stevens?"

"Of course, Captain. How else could my correspondents report to me with the efficiency you have observed?"

"Then why don't our receivers pick up the broadcasts of your agents? Dr. Wald says it's inherent in the principle that Dirac 'casts are picked up by *all* instruments tuned to receive them, bar none. And at this stage of the game there are so few such broadcasts being made that we'd be almost certain to detect any that weren't coming from our own operatives."

"I decline to answer that question, if you'll excuse the impoliteness," Stevens said, his voice quavering slightly. "I am an old man, Captain, and this intelligence agency is my sole source of income. If I told you how we operated, we would no longer have any advantage over your own service, except for the limited freedom from secrecy which we have. I have been assured by competent lawyers that I have every right to operate a private investigation bureau, properly licensed, upon any scale that I may choose; and that I have the right to keep my methods secret, as the so-called 'intellectual assets' of my firm. If you wish to use our services, well and good. We will provide them, with absolute guarantees on all information we furnish you, for an appropriate fee. But our methods are our own property."

Robin Weinbaum smiled twistedly. "I'm not a naïve man, Mr. Stevens," he said. "My service is hard on naïveté. You know as well as I do that the government can't allow you to operate on a free-lance basis, supplying top-secret information to anyone who can pay the price, or even free of charge to video columnists on a 'test' basis, even though you arrive at every jot of that information independently of espionage— which I still haven't entirely ruled out, by the way. If you can duplicate this *Brindisi* performance at will, we will have to have your services exclusively. In short, you become a hired civilian arm of my own bureau."

"Quite," Stevens said, returning the smile in a fatherly way. "We anticipated that, of course. However, we have contracts with other governments to consider; Erskine, in particular. If we are to work exclusively for Earth, necessarily our price will include compensation for renouncing our other accounts."

"Why should it? Patriotic public servants work for their government at a loss, if they can't work for it any other way."

"I am quite aware of that. I am quite prepared to renounce my other interests. But I do require to be paid."

"How much?" Weinbaum said, suddenly aware that his fists were clenched so tightly that they hurt.

Stevens appeared to consider, nodding his flowery white poll in senile deliberation. "My associates would have to be consulted. Tentatively, however, a sum equal to the present appropriation of your bureau would do, pending further negotiations."

Weinbaum shot to his feet, eyes wide. "You old buccaneer! You know damned well that I can't spend my entire appropriation on a single civilian service! Did it ever occur to you that most of the civilian outfits working for us are on cost-plus contracts, and that our civilian executives are being paid just a credit a year, by their own choice? You're demanding nearly two thousand credits an hour from your own government, and claiming the legal protection that the government affords you at the same time, in order to let those fanatics on Erskine run up a higher bid!"

"The price is not unreasonable," Stevens said. "The service is worth the price."

"That's where you're wrong! We have the discoverer of the machine working for us. For less than half the sum you're asking, we can find the application of the device that you're trading on—of that you can be damned sure."

"A dangerous gamble, Captain."

"Perhaps. We'll soon see!" Weinbaum glared at the placid face. "I'm forced to tell you that you're a free man, Mr. Stevens. We've been unable to show that you came by your information by any illegal method. You had classified facts in your possession, but no classified documents, and it's your privilege as a citizen to make guesses, no matter how educated.

"But we'll catch up with you sooner or later. Had you

been reasonable, you might have found yourself in a very good position with us, your income as assured as any political income can be, and your person respected to the hilt. Now, however, you're subject to censorship—you have no idea how humiliating that can be, but I'm going to see to it that you find out. There'll be no more newsbeats for Dana Lje, or for anyone else. I want to see every word of copy that you file with any client outside the bureau. Every word that is of use to me will be used, and you'll be paid the statutory one cent a word for it—the same rate that the FBI pays for anonymous gossip. Everything I don't find useful will be killed without clearance. Eventually we'll have the modification of the Dirac that you're using, and when that happens, you'll be so flat broke that a pancake with a harelip could spit right over you."

Weinbaum paused for a moment, astonished at his own fury.

Stevens's clarinetlike voice began to sound in the windowless cavity. "Captain, I have no doubt that you can do this to me, at least incompletely. But it will prove fruitless. I will give you a prediction, at no charge. It is guaranteed, as are all our predictions. It is this: *You will never find that modification.* Eventually, I will give it to you, on my own terms, but you will never find it for yourself, nor will you force it out of me. In the meantime, not a word of copy will be filed with you; for, despite the fact that you are an arm of the government, I can well afford to wait you out."

"Bluster," Weinbaum said.

"Fact. Yours is the bluster—loud talk based on nothing more than a hope. I, however, *know* whereof I speak. . . . But let us conclude this discussion. It serves no purpose; you will need to see my points made the hard way. Thank you for giving me my freedom. We will talk again under different circumstances on—let me see; ah, yes, on June 9 of the year 2091. That year is, I believe, almost upon us."

Stevens picked up his book again, nodding at Weinbaum, his expression harmless and kindly, his hands showing the marked tremor of *paralysis agitans.* Weinbaum moved helplessly to the door and flagged the turnkey. As the bars closed behind him, Stevens's voice called out: "Oh, yes; and a Happy New Year, Captain."

Weinbaum blasted his way back into his own office, at least twice as mad as the proverbial nest of hornets, and at the same time rather dismally aware of his own probable future. If Stevens's second prediction turned out to be as phenomenally accurate as his first had been, Capt. Robin Weinbaum would soon be peddling a natty set of second-hand uniforms.

He glared down at Margaret Soames, his receptionist. She glared right back; she had known him too long to be intimidated.

"Anything?" he said.

"Dr. Wald's waiting for you in your office. There are some field reports, and a couple of Diracs on your private tape. Any luck with the old codger?"

"That," he said crushingly, "is Top Secret."

"Poof. That means that nobody still knows the answer but J. Shelby Stevens."

He collapsed suddenly. "You're so right. That's just what it does mean. But we'll bust him wide open sooner or later. We've *got* to."

"You'll do it," Margaret said. "Anything else for me?"

"No. Tip off the clerical staff that there's a half holiday today, then go take in a stereo or a steak or something yourself. Dr. Wald and I have a few private wires to pull . . . and unless I'm sadly mistaken, a private bottle of aquavit to empty."

"Right," the receptionist said. "Tie one on for me, Chief. I understand that beer is the best chaser for aquavit—I'll have some sent up."

"If you should return after I am suitably squiffed," Weinbaum said, feeling a little better already, "I will kiss you for your thoughtfulness. *That* should keep you at your stereo at least twice through the third feature."

As he went on through the door of his own office, she said demurely behind him, "It certainly should."

As soon as the door closed, however, his mood became abruptly almost as black as before. Despite his comparative youth—he was now only fifty-five—he had been in the service a long time, and he needed no one to tell him the possible consequences which might flow from possession by a private citizen of the Dirac communicator. If there was ever to be a Federation of Man in the Galaxy, it was within the power of J. Shelby Stevens to ruin it before it had fairly gotten started. And there seemed to be nothing at all that could be done about it.

"Hello, Thor," he said glumly. "Pass the bottle."

"Hello, Robin. I gather things went badly. Tell me about it."

Briefly, Weinbaum told him. "And the worst of it," he finished, "is that Stevens himself predicts that we won't find the application of the Dirac that he's using, and that eventually we'll have to buy it at his price. Somehow I believe him—but I can't see how it's possible. If I were to tell Congress that I was going to spend my entire appropriation for a single civilian service, I'd be out on my ear within the next three sessions."

"Perhaps that isn't his real price," the scientist suggested. "If he wants to barter, he'd naturally begin with a demand miles above what he actually wants."

"Sure, sure . . . but frankly, Thor, I'd hate to give the old reprobate even a single credit if I could get out of it." Weinbaum sighed. "Well, let's see what's come in from the field."

Thor Wald moved silently away from Weinbaum's desk while the officer unfolded it and set up the Dirac screen. Stacked neatly next to the ultraphone—a device Weinbaum had been thinking of, only a few days ago, as permanently outmoded—were the tapes Margaret had mentioned. He fed the first one into the Dirac and turned the main toggle to the position labeled START.

Immediately the whole screen went pure white and the audio speakers emitted an almost instantly end-stopped blare of sound—a *beep* which, as Weinbaum already knew, made up a continuous spectrum from about 30 cycles per second to well above 18,000 cps. Then both the light and the noise were gone as if they had never been, and were replaced by the familiar face and voice of Weinbaum's local ops chief in Rico City.

"There's nothing unusual in the way of transmitters in Stevens's offices here," the operative said without preamble. "And there isn't any local Interstellar Information staff, except for one stenographer, and she's as dumb as they come. About all we could get from her is that Stevens is 'such a sweet old man.' No possibility that she's faking it; she's genuinely stupid, the kind that thinks Betelgeuse is something Indians use to darken their skins. We looked for some sort of list or code table that would give us a line on Stevens's field staff, but that was another dead end. Now we're maintaining a twenty-four-hour Dinwiddie watch on the place from a joint across the street. Orders?"

Weinbaum dictated to the blank stretch of tape which
followed: "Margaret, next time you send any Dirac tapes
in here, cut that damnable *beep* off them first. Tell the boys
in Rico City that Stevens has been released, and that I'm
proceeding for an Order In Security to tap his ultraphone
and his local lines—this is one case where I'm sure we can
persuade the court that tapping's necessary. Also—and be
damned sure you code this—tell them to proceed with the
tap immediately and to maintain it regardless of whether
or not the court O.K.s it. I'll thumbprint a Full Responsibility Confession for them. We can't afford to play pat-
a-cake with Stevens—the potential is just too damned big.
And oh, yes, Margaret, send the message by carrier, and
send out general orders to everybody concerned not to use
the Dirac again except when distance and time rule every
other medium out. Stevens has already admitted that he can
receive Dirac 'casts."

He put down the mike and stared morosely for a moment
at the beautiful Eridanean scrollwood of his desktop. Wald
coughed inquiringly and retrieved the aquavit.

"Excuse me, Robin," he said, "but I should think that would
work both ways."

"So should I. And yet the fact is that we've never picked
up so much as a whisper from either Stevens or his agents.
I can't think of any way that could be pulled, but evidently
it can."

"Well, let's rethink the problem, and see what we get,"
Wald said. "I didn't want to say so in front of the young
lady, for obvious reasons—I mean Miss Lje, of course, not
Margaret—but the truth is that the Dirac is essentially a
simple mechanism in principle. I seriously doubt that there's
any way to transmit a message from it which can't be
detected—and an examination of the theory with that proviso
in mind might give us something new."

"What proviso?" Weinbaum said. Thor Wald left him
behind rather often these days.

"Why, that a Dirac transmission doesn't *necessarily* go to
all communicators capable of receiving it. If that's true,
then the reasons why it is true should emerge from the
theory."

"I see. O.K., proceed on that line. I've been looking at
Stevens's dossier while you were talking, and it's an absolute
desert. Prior to the opening of the office in Rico City, there's
no dope whatever on J. Shelby Stevens. The man as good

as rubbed my nose in the fact that he's using a pseud when I first talked to him. I asked him what the 'J' in his name stood for, and he said, 'Oh, let's make it Jerome.' But who the man behind the pseud *is* . . ."

"Is it possible that he's using his own initials?"

"No," Weinbaum said. "Only the dumbest ever do that, or transpose syllables, or retain any connection at all with their real names. Those are the people who are in serious emotional trouble, people who drive themselves into anonymity, but leave clues strewn all around the landscape—those clues are really a cry for help, for discovery. Of course we're working on that angle—we can't neglect anything—but J. Shelby Stevens isn't that kind of case, I'm sure." Weinbaum stood up abruptly. "O.K., Thor—what's first on your technical program?"

"Well . . . I suppose we'll have to start with checking the frequencies we use. We're going on Dirac's assumption—and it works very well, and always has—that a positron in motion through a crystal lattice is accompanied by de Broglie waves which are transforms of the waves of an electron in motion somewhere else in the Universe. Thus if we control the frequency and path of the positron, we control the placement of the electron—we cause it to appear, so to speak, in the circuits of a communicator somewhere else. After that, reception is just a matter of amplifying the bursts and reading the signal."

Wald scowled and shook his blond head. "If Stevens is getting out messages which we don't pick up, my first assumption would be that he's worked out a fine-tuning circuit that's more delicate than ours, and is more or less sneaking his messages under ours. The only way that could be done, as far as I can see at the moment, is by something really fantastic in the way of exact frequency control of his positron gun. If so, the logical step for us is to go back to the beginning of our tests and rerun our diffractions to see if we can refine our measurements of positron frequencies."

The scientist looked so inexpressibly gloomy as he offered this conclusion that a pall of hopelessness settled over Weinbaum in sheer sympathy. "You don't look as if you expected that to uncover anything new."

"I don't. You see, Robin, things are different in physics now than they used to be in the twentieth century. In those days, it was always presupposed that physics was limitless—the classic statement was made by Weyl, who said that

'It is the nature of a real thing to be inexhaustible in content.' We know now that that's not so, except in a remote, associational sort of way. Nowadays, physics is a defined and self-limited science; its scope is still prodigious, but we can no longer think of it as endless.

"This is better established in particle physics than in any other branch of the science. Half of the trouble physicists of the last century had with Euclidean geometry—and hence the reason why they evolved so many recomplicated theories of relativity—is that it's a geometry of lines, and thus can be subdivided infinitely. When Cantor proved that there really is an infinity, at least mathematically speaking, that seemed to clinch the case for the possibility of a really infinite physical universe, too."

Wald's eyes grew vague, and he paused to gulp down a slug of the licorice-flavored aquavit which would have made Weinbaum's every hair stand on end.

"I remember," Wald said, "the man who taught me theory of sets at Princeton, many years ago. He used to say: 'Cantor teaches us that there are many kinds of infinities. *There* was a crazy old man!' "

Weinbaum rescued the bottle hastily. "So go on, Thor."

"Oh." Wald blinked. "Yes. Well, what we know now is that the geometry which applies to ultimate particles, like the positron, isn't Euclidean at all. It's Pythagorean—a geometry of points, not lines. Once you've measured one of those points, and it doesn't matter what kind of quantity you're measuring, you're down as far as you can go. At that point, the Universe becomes discontinuous, and no further refinement is possible.

"And I'd say that our positron-frequency measurements have already gotten that far down. There isn't another element in the Universe denser than plutonium, yet we get the same frequency values by diffraction through plutonium crystals that we get through osmium crystals—there's not the slightest difference. If J. Shelby Stevens is operating in terms of fractions of those values, then he's doing what an organist would call 'playing in the cracks'—which is certainly something you can *think* about doing, but something that's in actuality impossible to do. *Hoop*."

"Hoop?" Weinbaum said.

"Sorry. A hiccup only."

"Oh. Well, maybe Stevens has rebuilt the organ?"

"If he has rebuilt the metrical frame of the Universe to

accommodate a private skip-tracing firm," Wald said firmly, "I for one see no reason why we can't countercheck him —hoop—by declaring the whole cosmos null and void."

"All right, all right," Weinbaum said, grinning. "I didn't mean to push your analogy right over the edge—I was just asking. But let's get to work on it anyhow. We can't just sit here and let Stevens get away with it. If this frequency angle turns out to be as hopeless as it seems, we'll try something else."

Wald eyed the aquavit bottle owlishly. "It's a very pretty problem," he said. "Have I ever sung you the song we have in Sweden called 'Nat-og-Dag?' "

"*Hoop*," Weinbaum said, to his own surprise, in a high falsetto. "Excuse me. No. Let's hear it."

The computer occupied an entire floor of the Security building, its seemingly identical banks laid out side by side on the floor along an advanced pathological state of Peano's "space-filling curve." At the current business end of the line was a master control board with a large television screen at its center, at which Dr. Wald was stationed, with Weinbaum looking, silently but anxiously, over his shoulder.

The screen itself showed a pattern which, except that it was drawn in green light against a dark gray background, strongly resembled the grain in a piece of highly polished mahogany. Photographs of similar patterns were stacked on a small table to Dr. Wald's right; several had spilled over onto the floor.

"Well, there it is," Wald sighed at length. "And I won't struggle to keep myself from saying 'I told you so.' What you've had me do here, Robin, is to reconfirm about half the basic postulates of particle physics—which is why it took so long, even though it was the first project we started." He snapped off the screen. "There are no cracks for J. Shelby to play in. That's definite."

"If you'd said 'That's flat,' you would have made a joke," Weinbaum said sourly. "Look . . . isn't there still a chance of error? If not on your part, Thor, then in the computer? After all, it's set up to work only with the unit charges of modern physics; mightn't we have to disconnect the banks that contain that bias before the machine will follow the fractional-charge instructions we give it?"

" 'Disconnect,' he says," Wald groaned, mopping his brow reflectively. "The bias exists everywhere in the machine, my friend, because it functions everywhere on those same unit charges. It wasn't a matter of subtracting banks; we had to add one with a bias all its own, to countercorrect the corrections the computer would otherwise apply to the instructions. The technicians thought I was crazy. Now, five months later, I've proved it."

Weinbaum grinned in spite of himself. "What about the other projects?"

"All done—some time back, as a matter of fact. The staff and I checked every single Dirac tape we've received since you released J. Shelby from Yaphank, for any sign of intermodulation, marginal signals, or anything else of the kind. There's nothing, Robin, absolutely nothing. That's our net result, all around."

"Which leaves us just where we started," Weinbaum said. "All the monitoring projects came to the same dead end; I strongly suspect that Stevens hasn't risked any further calls from his home office to his field staff, even though he seemed confident that we'd never intercept such calls—as we haven't. Even our local wire tapping hasn't turned up anything but calls by Stevens's secretary, making appointments for him with various clients, actual and potential. Any information he's selling these days he's passing on in person—and not in his office, either, because we've got bugs planted all over that and haven't heard a thing."

"That must limit his range of operation enormously," Wald objected.

Weinbaum nodded. "Without a doubt—but he shows no signs of being bothered by it. He can't have sent any tips to Erskine recently, for instance, because our last tangle with that crew came out very well for us, even though we had to use the Dirac to send the orders to our squadron out there. If he overheard us, he didn't even try to pass the word. Just as he said, he's sweating us out—" Weinbaum paused. "Wait a minute, here comes Margaret. And by the length of her stride, I'd say she's got something particularly nasty on her mind."

"You bet I do," Margaret Soames said vindictively. "And it'll blow plenty of lids around here, or I miss my guess. The I. D. squad has finally pinned down J. Shelby Stevens. They did it with the voice-comparator alone."

"How does that work?" Wald said interestedly.

"Blink microphone," Weinbaum said impatiently. "Isolates inflections on single, normally stressed syllables and matches them. Standard I. D. searching technique, on a case of this kind, but it takes so long that we usually get the quarry by other means before it pays off. Well, don't stand there like a dummy, Margaret. Who is he?

" 'He,' " Margaret said, "is your sweetheart of the video waves, Miss Dana Lje."

"They're crazy!" Wald said, staring at her.

Weinbaum came slowly out of his first shock of stunned disbelief. "No, Thor," he said finally. "No, it figures. If a woman is going to go in for disguises, there are always two she can assume outside her own sex: a young boy, and a very old man. And Dana's an actress; that's no news to us."

"But—but why did she do it, Robin?"

"That's what we're going to find out right now. So we wouldn't get the Dirac modification by ourselves, eh! Well, there are other ways of getting answers besides particle physics. Margaret, do you have a pick-up order out for that girl?"

"No," the receptionist said. "This is one chestnut I wanted to see you pull out for yourself. You give me the authority, and I send the order—not before."

"Spiteful child. Send it, then, and glory in my gritted teeth. Come on, Thor—let's put the nutcracker on this chestnut."

As they were leaving the computer floor, Weinbaum stopped suddenly in his tracks and began to mutter in an almost inaudible voice.

Wald said, "What's the matter, Robin?"

"Nothing. I keep being brought up short by those predictions. What's the date?"

"M'm . . . June 9. Why?"

"It's the exact date that 'Stevens' predicted we'd meet again, damn it! Something tells me that this isn't going to be as simple as it looks."

If Dana Lje had any idea of what she was in for—and considering the fact that she was 'J. Shelby Stevens' it had to be assumed that she did—the knowledge seemed not to make her at all fearful. She sat as composedly as ever before Weinbaum's desk, smoking her eternal cigarette, and waited, one dimpled knee pointed directly at the bridge of the officer's nose.

"Dana," Weinbaum said, "this time we're going to get all the answers, and we're not going to be gentle about it. Just in case you're not aware of the fact, there are certain laws relating to giving false information to a security officer, under which we could heave you in prison for a minimum of fifteen years. By application of the statutes on using communications to defraud, plus various local laws against transvestism, pseudonymity and so on, we could probably pile up enough additional short sentences to keep you in Yaphank until you really *do* grow a beard. So I'd advise you to open up."

"I have every intention of opening up," Dana said. "I know, practically word for word, how this interview is going to proceed, what information I'm going to give you, just when I'm going to give it to you—and what you're going to pay me for it. I knew all that many months ago. So there would be no point in my holding out on you."

"What you're saying, Miss Lje," Thor Wald said in a re-signed voice, "is that the future is fixed, and that you can read it, in every essential detail."

"Quite right, Dr. Wald. Both those things are true."

There was a brief silence.

"All right," Weinbaum said grimly. "Talk."

"All right, Captain Weinbaum, pay me," Dana said calmly. Weinbaum snorted.

"But I'm quite serious," she said. "You still don't know what I know about the Dirac communicator. I won't be forced to tell it, by threat of prison or by any other threat. You see, I know for a fact that you aren't going to send me to prison, or give me drugs, or do anything else of that kind. I know for a fact, instead, that you are going to pay me—so I'd be very foolish to say a word until you do. After all, it's quite a secret you're buying. Once I tell you what it is, you and the entire service will be able to read the future as I do, and then the information will be valueless to me."

Weinbaum was completely speechless for a moment. Finally he said, "Dana, you have a heart of purest brass, as well as a knee with an invisible gunsight on it. I say that I'm *not* going to give you my appropriation, regardless of what the future may or may not say about it. I'm not going to give it to you because the way my government—and yours —runs things makes such a price impossible. Or is that really your price?"

"It's my real price . . . but it's also an alternative. Call it my second choice. My first choice, which means the price I'd settle for, comes in two parts: (a) to be taken into your service as a responsible officer; and, (b) to be married to Captain Robin Weinbaum."

Weinbaum sailed up out of his chair. He felt as though copper-colored flames a foot long were shooting out of each of his ears.

"Of all the—" he began. There his voice failed completely.

From behind him, where Wald was standing, came something like a large, Scandinavian-model guffaw being choked into insensibility.

Dana herself seemed to be smiling a little.

"You see," she said, "I don't point my best and most accurate knee at every man I meet."

Weinbaum sat down again, slowly and carefully. "Walk, do not run, to nearest exit," he said. "Women and childlike security officers first. Miss Lje, are you trying to sell me the notion that you went through this elaborate hanky-panky—beard and all—out of a burning passion for my dumpy and underpaid person?"

"Not entirely," Dana Lje said. "I want to be in the bureau, too, as I said. Let me confront you, though, Captain, with a fact of life that doesn't seem to have occurred to you at all. Do you accept as a fact that I can read the future in detail, and that that, to be possible at all, means that the future is fixed?"

"Since Thor seems able to accept it, I suppose I can too—provisionally."

"There's nothing provisional about it," Dana said firmly. "Now, when I first came upon this—uh, this gimmick—quite a while back, one of the first things that I found out was that I was going to go through the 'J. Shelby Stevens' masquerade, force myself onto the staff of the bureau, and marry you, Robin. At the time, I was both astonished and completely rebellious. I didn't want to be on the bureau staff; I liked my free-lance life as a video commentator. I didn't want to marry you, though I wouldn't have been averse to living with you for a while—say a month or so. And above all, the masquerade struck me as ridiculous.

"But the facts kept staring me in the face. I *was* going to do all those things. There were no alternatives, no fanciful 'branches of time,' no decision-points that might be altered to make the future change. My future, like yours, Dr. Wald's, and everyone else's, was fixed. It didn't matter a snap whether or not I had a decent motive for what I was going to do; I was going to do it anyhow. Cause and effect, as I could see for myself, just don't exist. One event follows an-

other because events are just as indestructible in space-time
as matter and energy are.

"It was the bitterest of all pills. It will take me many
years to swallow it completely, and you too. Dr. Wald will
come around a little sooner, I think. At any rate, once I was
intellectually convinced that all this was so, I had to protect
my own sanity. I knew that I couldn't alter what I was going
to do, but the least I could do to protect myself was to supply
myself with motives. Or, in other words, just plain rationalizations. That much, it seems, we're free to do; the consciousness of the observer is just along for the ride through time,
and can't alter events—but it can comment, explain, invent.
That's fortunate, for none of us could stand going through
motions which were truly free of what we think of as
personal significances.

"So I supplied myself with the obvious motives. Since I
was going to be married to you and couldn't get out of it, I
set out to convince myself that I loved you. Now I do. Since
I was going to join the bureau staff, I thought over all the
advantages that it might have over video commentating, and
found that they made a respectable list. Those are my
motives.

"But I had no such motives at the beginning. Actually,
there are never motives behind actions. All actions are fixed.
What we called motives evidently are rationalizations by the
helpless observing consciousness, which is intelligent enough
to smell an event coming—and, since it cannot avert the
event, instead cooks up reasons for wanting it to happen."

"Wow," Dr. Wald said, inelegantly but with considerable
force.

"Either 'wow' or 'balderdash' seems to be called for—I
can't quite decide which," Weinbaum agreed. "We know that
Dana is an actress, Thor, so let's not fall off the apple tree
quite yet. Dana, I've been saving the *really* hard question for
the last. That question is: *How?* How did you arrive at this
modification of the Dirac transmitter? Remember, we know
your background, where we didn't know that of 'J. Shelby
Stevens.' You're not a scientist. There were some fairly high-
powered intellects among your distant relatives, but that's as
close as you come."

"I'm going to give you several answers to that question,"
Dana Lje said. "Pick the one you like best. They're all true,
but they tend to contradict each other here and there.

"To begin with, you're right about my relatives, of course.

If you'll check your dossier again, though, you'll discover that those so-called 'distant' relatives were the last surviving members of my family besides myself. When they died, second and fourth and ninth cousins though they were, their estates reverted to me, and among their effects I found a sketch of a possible instantaneous communicator based on de Broglie-wave inversion. The material was in very rough form, and mostly beyond my comprehension, because I am, as you say, no scientist myself. But I was interested; I could see, dimly, what such a thing might be worth—and not only in money.

"My interest was fanned by two coincidences—the kind of coincidences that cause-and-effect just can't allow, but which seem to happen all the same in the world of unchangeable events. For most of my adult life, I've been in communications industries of one kind or another, mostly branches of video. I had communications equipment around me constantly, and I had coffee and doughnuts with communications engineers every day. First I picked up the jargon; then, some of the procedures; and eventually a little real knowledge. Some of the things I learned can't be gotten any other way. Some other things are ordinarily available only to highly educated people like Dr. Wald here, and came to me by accident, in horseplay, between kisses, and a hundred other ways—all natural to the environment of a video network."

Weinbaum found, to his own astonishment, that the "between kisses" clause did not sit very well in his chest. He said, with unintentional brusqueness: "What's the other coincidence?"

"A leak in your own staff."

"Dana, you ought to have that set to music."

"Suit yourself."

"I can't suit myself," Weinbaum said petulantly. "I work for the government. Was this leak direct to you?"

"Not at first. That was why I kept insisting to you in person that there might be such a leak, and why I finally began to hint about it in public, on my program. I was hoping that you'd be able to seal it up inside the bureau before my first rather tenuous contact with it got lost. When I didn't succeed in provoking you into protecting yourself, I took the risk of making direct contact with the leak myself—and the first piece of secret information that came to me through it was the final point I needed to put my Dirac communicator together. When it was all assembled, it did more than just communicate. It predicted. And I can tell you why."

Weinbaum said thoughtfully, "I don't find this very hard to accept, so far. Pruned of the philosophy, it even makes some sense of the 'J. Shelby Stevens' affair. I assume that by letting the old gentleman become known as somebody who knew more about the Dirac transmitter than I did, and who wasn't averse to negotiating with anybody who had money, you kept the leak working through you—rather than transmitting data directly to unfriendly governments."

"It did work out that way," Dana said. "But that wasn't the genesis or the purpose of the Stevens masquerade. I've already given you the whole explanation of how that came about."

"Well, you'd better name me that leak, before the man gets away."

"When the price is paid, not before. It's too late to prevent a getaway, anyhow. In the meantime, Robin, I want to go on and tell you the other answer to your question about how I was able to find this particular Dirac secret, and you didn't. What answers I've given you up to now have been cause-and-effect answers, with which we're all more comfortable. But I want to impress on you that all apparent cause-and-effect relationships are accidents. There is no such thing as a cause, and no such thing as an effect. I found the secret because I found it; that event was fixed; that certain circumstances seem to explain why I found it, in the old cause-and-effect terms, is irrelevant. Similarly, with all your superior equipment and brains, you didn't find it for one reason, and one reason alone: because you didn't find it. The history of the future says you didn't."

"I pays my money and I takes no choice, eh?" Weinbaum said ruefully.

"I'm afraid so—and I don't like it any better than you do."

"Thor, what's your opinion of all this?"

"It's just faintly flabbergasting," Wald said soberly. "However, it hangs together. The deterministic universe which Miss Lje paints was a common feature of the old relativity theories, and as sheer speculation has an even longer history. I would say that, in the long run, how much credence we place in the story as a whole will rest upon her method of, as she calls it, reading the future. If it is demonstrable beyond any doubt, then the rest becomes perfectly credible—philosophy and all. If it doesn't, then what remains is an admirable job of acting, plus some metaphysics which, while self-consistent, is not original with Miss Lje."

"That sums up the case as well as if I'd coached you, Dr. Wald," Dana said. "I'd like to point out one more thing. If I can read the future, then 'J. Shelby Stevens' never had any need for a staff of field operatives, and he never needed to send a single Dirac message which you might intercept. All he needed to do was to make predictions from his readings, which he knew to be infallible; no private espionage network had to be involved."

"I see that," Weinbaum said dryly. "All right, Dana, let's put the proposition this way: *I do not believe you.* Much of what you say is probably true, but in totality I believe it to be false. On the other hand, if you're telling the whole truth, you certainly deserve a *place* on the bureau staff—it would be dangerous as hell *not* to have you with us—and the marriage is a more or less minor matter, except to you and me. You can have that with no strings attached; I don't want to be bought, any more than you would.

"So: if you will tell me where the leak is, we will consider that part of the question closed. I make that condition not as a price, but because I don't want to get myself engaged to somebody who might be shot as a spy within a month."

"Fair enough," Dana said. "Robin, your leak is Margaret Soames. She is an Erskine operative, and nobody's bubble-brain. She's a highly trained technician."

"Well, I'll be damned," Weinbaum said in astonishment. "Then she's already flown the coop—she was the one who first told me we'd identified you. She must have taken on that job in order to hold up delivery long enough to stage an exit."

"That's right. But you'll catch her, day after tomorrow. And you are now a hooked fish, Robin."

There was another suppressed burble from Thor Wald.

"I accept the fate happily," Weinbaum said, eying the gunsight knee. "Now, if you will tell me how you work your swami trick, and if it backs up everything you've said to the letter, as you claim, I'll see to it that you're also taken into the bureau and that all charges against you are quashed. Otherwise, I'll probably have to kiss the bride between the bars of a cell."

Dana smiled. "The secret is very simple. It's in the beep."

Weinbaum's jaw dropped. "The beep? The Dirac noise?"

"That's right. You didn't find it out because you considered the beep to be just a nuisance, and ordered Miss Soames to cut it off all tapes before sending them in to you. Miss Soames, who had some inkling of what the beep meant, was more than happy to do so, leaving the reading of the beep exclusively to 'J. Shelby Stevens'—who she thought was going to take on Erskine as a client."

"Explain," Thor Wald said, looking intense.

"Just as you assumed, every Dirac message that is sent is picked up by every receiver that is capable of detecting it. *Every* receiver—including the first one ever built, which is yours, Dr. Wald, through the hundreds of thousands of them which will exist throughout the Galaxy in the twenty-fourth century, to the untold millions which will exist in the thirtieth century, and so on. The Dirac beep is the simultaneous reception of *every one of the Dirac messages which have ever been sent, or ever will be sent.* Incidentally, the cardinal number of the total of those messages is a relatively small and of course finite number; it's far below really large finite numbers such as the number of electrons in the universe, even when you break each and every message down into individual 'bits' and count those."

"Of course," Dr. Wald said softly. "Of course! But, Miss Lje . . . how do you tune for an individual message? We tried fractional positron frequencies, and got nowhere."

"I didn't even know fractional positron frequencies existed," Dana confessed. "No, it's simple—so simple that a lucky layman like me could arrive at it. You tune individual messages out of the beep by time lag, nothing more. All the messages arrive at the same instant, in the smallest fraction of time that exists, something called a 'chronon.' "

"Yes," Wald said. "The time it takes one electron to move from one quantum-level to another. That's the Pythagorean point of time measurement."

"Thank you. Obviously no gross physical receiver can respond to a message that brief, or at least that's what I thought at first. But because there are relay and switching delays, various forms of feedback and so on, in the apparatus itself, the beep arrives at the output end as a complex pulse which has been 'splattered' along the time axis for a full second or more. That's an effect which you can exaggerate by recording the 'splattered' beep on a high-speed tape, the same way you would record any event that you wanted to study in slow motion. Then you tune up the various failure-points in your receiver, to exaggerate one failure, minimize all the others, and use noise-suppressing techniques to cut out the background."

Thor Wald frowned. "You'd still have a considerable garble when you were through. You'd have to sample the messages——"

"Which is just what I did; Robin's little lecture to me about the ultrawave gave me that hint. I set myself to find out how the ultrawave channel carries so many messages at once, and I discovered that you people sample the incoming pulses every thousandth of a second and pass on one pip only when the wave deviates in a certain way from the mean. I didn't really believe it would work on the Dirac beep, but it turned out just as well: 90 percent as intelligible as the original transmission after it came through the smearing device. I'd already got enough from the beep to put my plan in motion, of course—but now every voice message in it was available, and crystal-clear: If you select three pips every thousandth of second, you can even pick up an intelligible transmission of music—a little razzy, but good enough to identify the instruments that are playing—and that's a very close test of any communications device."

"There's a question of detail here that doesn't quite follow," said Weinbaum, for whom the technical talk was becoming a little too thick to fight through. "Dana, you say that you knew the course this conversation was going to take—yet it isn't being Dirac-recorded, nor can I see any reason why any summary of it would be sent out on the Dirac afterwards."

"That's true, Robin. However, when I leave here, I will make such a transcast myself, on my own Dirac. Obviously I will—because I've *already* picked it up, from the beep."

"In other words, you're going to call yourself up—months ago."

"That's it," Dana said. "It's not as useful a technique as you might think at first, because it's dangerous to make such broadcasts while a situation is still developing. You can safely 'phone back' details only after the given situation has gone to completion, as a chemist might put it. Once you know, however, that when you use the Dirac you're dealing with time, you can coax some very strange things out of the instrument."

She paused and smiled. "I have heard," she said conversationally, "the voice of the President of our Galaxy, in 3480, announcing the federation of the Milky Way and the

Magellanic Clouds. I've heard the commander of a world-line cruiser, traveling from 8873 to 8704 along the world line of the planet Hathshepa, which circles a star on the rim of NGC 4725, calling for help across eleven million light-years—but what kind of help he was calling for, or will be calling for, is beyond my comprehension. And many other things. When you check on me, you'll hear these things too—and you'll wonder what many of them mean.

"And you'll listen to them even more closely than I did, in the hope of finding out whether or not anyone was able to understand in time to help."

Weinbaum and Wald looked dazed.

Her voice became a little more somber. "Most of the voices in the Dirac beep are like that—they're cries for help, which you can overhear decades or centuries before the senders get into trouble. You'll feel obligated to answer every one, to try to supply the held that's needed. And you'll listen to the succeeding messages and say: 'Did we—will we get there in time? Did we understand in time?'

"And in most cases you won't be sure. You'll know the future, but not what most of it means. The farther into the future you travel with the machine, the more incomprehensible the messages become, and so you're reduced to telling yourself that time will, after all, have to pass by at its own pace, before enough of the surrounding events can emerge to make those remote messages clear.

"The long-run effect, as far as I can think it through, is not going to be that of omniscience—of our consciousness being extracted entirely from the time stream and allowed to view its whole sweep from one side. Instead, the Dirac in effect simply slides the bead of consciousness forward from the present a certain distance. Whether it's five hundred or five thousand years still remains to be seen. At that point the law of diminishing returns sets in—or the noise factor begins to overbalance the information, take your choice—and the observer is reduced to traveling in time at the same old speed. He's just a bit ahead of himself."

"You've thought a great deal about this," Wald said slowly. "I dislike to think of what might have happened had some less conscientious person stumbled on the beep."

"That wasn't in the cards," Dana said.

In the ensuing quiet, Weinbaum felt a faint, irrational sense of let-down, of something which had promised more than had been delivered—rather like the taste of fresh bread as compared to its smell, or the discovery that Thor Wald's Swedish "folk song" *Nat-og-Dag* was only Cole Porter's *Night and Day* in another language. He recognized the feeling: it was the usual emotion of the hunter when the hunt is over, the born detective's professional version of the *post coitum tristre*. After looking at the smiling, supple Dana Lje a moment more, however, he was almost content.

"There's one more thing," he said. "I don't want to be insufferably skeptical about this—but I want to see it work. Thor, can we set up a sampling and smearing device such as Dana describes and run a test?"

"In fifteen minutes," Dr. Wald said. "We have most of the unit in already assembled form on our big ultrawave receiver, and it shouldn't take any effort to add a high-speed tape unit to it. I'll do it right now."

He went out. Weinbaum and Dana looked at each other for a moment, rather like strange cats. Then the security officer got up, with what he knew to be an air of somewhat grim determination, and seized his fiancée's hands, anticipating a struggle.

That first kiss was, by intention at least, mostly *pro forma*. But by the time Wald padded back into the office, the letter had been pretty thoroughly superseded by the spirit. The scientist harrumphed and set his burden on the desk. "This is all there is to it," he said, "but I had to hunt all through the library to find a Dirac record with a beep still on it. Just a moment more while I make connections. . . ."

Weinbaum used the time to bring his mind back to the matter at hand, although not quite completely. Then two tape spindles began to whir like so many bees, and the end-stopped sound of the Dirac beep filled the room. Wald stopped the apparatus, reset it, and started the smearing tape very slowly in the opposite direction.

A distant babble of voices came from the speaker. As Weinbaum leaned forward tensely, one voice said clearly and loudly above the rest:

"Hello, Earth bureau. Lt. T. L. Matthews at Hercules Station NGC 6341, transmission date 13-22-2091. We have the last point on the orbit curve of your dope-runners plotted, and the curve itself points to a small system about twenty-five light-years from the base here; the place hasn't even got a name on our charts. Scouts show the home planet at least twice as heavily fortified as we anticipated, so we'll need another cruiser. We have a 'can-do' from you in the beep for us, but we're waiting as ordered to get it in the present. NGC 6341 Matthews out."

After the first instant of stunned amazement—for no amount of intellectual willingness to accept could have prepared him for the overwhelming fact itself—Weinbaum had grabbed a pencil and begun to write at top speed. As the voice signed out he threw the pencil down and looked excitedly at Dr. Wald.

"Seven months ahead," he said, aware that he was grinning like an idiot. "Thor, you know the trouble we've had with that needle in the Hercules haystack! This orbit-curve trick must be something Matthews has yet to dream up—at least he hasn't come to me with it yet, and there's nothing in the situation as it stands now that would indicate a closing time of six months for the case. The computers said it would take three more years."

"It's new data," Dr. Wald agreed solemnly.

"Well, don't stop there, in God's name! Let's hear some more!"

Dr. Wald went through the ritual, much faster this time. The speaker said:

"Nausentampen. Eddettompic. Berobsilom. Aimkaksetchoc. Sanbetogmow. Datdectamset. Domatrosmin. Out."

"My word," Wald said. "What's all that?"

"That's what I was talking about," Dana Lje said. "At least half of what you get from the beep is just as incomprehensible. I suppose it's whatever has happened to the English language, thousands of years from now."

"No, it isn't," Weinbaum said. He had resumed writing, and was still at it, despite the comparative briefness of the transmission. "Not this sample, anyhow. That, ladies and gentlemen, is code—no language consists exclusively of four-syllable words, of that you can be sure. What's more, it's a version of our code. I can't break it down very far— it takes a full-time expert to read this stuff—but I get the date and some of the sense. It's March 12, 3022, and there's some kind of a mass evacuation taking place. The message seems to be a routing order."

"But why will we be using code?" Dr. Wald wanted to know. "It implies that we think somebody might overhear us—somebody else with a Dirac. That could be very messy."

"It could indeed," Weinbaum said. "But we'll find out, I imagine. Give her another spin, Thor."

"Shall I try for a picture this time?"

Weinbaum nodded. A moment later, he was looking squarely into the green-skinned face of something that looked like an animated traffic signal with a helmet on it. Though the creature had no mouth, the Dirac speaker was saying quite clearly, "Hello, Chief. This is Thammos NGC 2287, transmission date Gor 60, 302 by my calendar, July 2, 2973 by yours. This is a lousy little planet. Everything stinks of oxygen, just like Earth. But the natives accept us and that's the important thing. We've got your genius safely born. Detailed report coming later by paw. NGC 2287 Thammos out."

"I wish I knew my New General Catalogue better," Weinbaum said. "Isn't that M 41 in Canis Major, the one with the red star in the middle? And we'll be using non-humanoids there! What *was* that creature, anyhow? Never mind, spin her again."

Dr. Wald spun her again. Weinbaum, already feeling a little dizzy, had given up taking notes. That could come later, all that could come later. Now he wanted only scenes and voices, more and more scenes and voices from the future. They were better than aquavit, even with a beer chaser.

### 4

THE INDOCTRINATION tape ended, and Krasna touched a button. The Dirac screen darkened, and folded silently back into the desk.

"They didn't see their way through to us, not by a long shot," he said. "They didn't see, for instance, that when one section of the government becomes nearly all-knowing—no matter how small it was to begin with—it necessarily becomes all of the government that there is. Thus the bureau turned into the Service and pushed everyone else out.

"On the other hand, those people did come to be afraid that a government with an all-knowing arm might become a rigid dictatorship. That couldn't happen and didn't happen, because the more you know, the wider your field of possible operation becomes and the more fluid and dynamic a society you need. How could a rigid society expand to other star systems, let alone other galaxies? It couldn't be done."

"I should think it could," Jo said slowly. "After all, if you know in advance what everybody is going to do . . ."

"But we don't, Jo. That's just a popular fiction—or, if

you like, a red herring. Not all of the business of the cosmos is carried on over the Dirac, after all. The only events we can ever overhear are those which are transmitted as a message. Do you order your lunch over the Dirac? Of course you don't. Up to now, you've never said a word over the Dirac in your life.

"And there's much more to it than that. All dictatorships are based on the proposition that government can somehow control a man's thoughts. We know now that the consciousness of the observer is the only free thing in the Universe. Wouldn't we look foolish trying to control that, when our entire physics shows that it's impossible to do so? That's why the Service is in no sense a thought police. We're interested only in acts. We're an Event Police."

"But why?" Jo said. "If all history is fixed, why do we bother with these boy-meets-girl assignments, for instance? The meetings will happen anyhow."

"Of course they will," Krasna agreed immediately. "But look, Jo. Our interests as a government depend upon the future. We operate *as if* the future is as real as the past, and so far we haven't been disappointed: the Service is 100 per cent successful. But that very success isn't without its warnings. What would happen if we *stopped* supervising events? We don't know, and we don't dare take the chance. Despite the evidence that the future is fixed, we have to take on the role of the caretaker of inevitability. We believe that nothing can possibly go wrong . . . but we have to act on the philosophy that history helps only those who help themselves.

"That's why we safeguard huge numbers of courtships right through to contract, and even beyond it. We have to see to it that *every single person who is mentioned in any Dirac 'cast gets born.* Our obligation as Event Police is to make the events of the future possible, because those events are crucial to our society—even the smallest of them. It's an enormous task, believe me, and it gets bigger and bigger every day. Apparently it always will."

"Always?" Jo said. "What about the public? Isn't it going to smell this out sooner or later? The evidence is piling up at a terrific rate."

"Yes and no," Krasna said. "Lots of people are smelling it out right now, just as you did. But the number of new people we need in the Service grows faster—it's always ahead of the number of laymen who follow the clues to the truth."

Jo took a deep breath. "You take all this as if it were as commonplace as boiling an egg, Kras," he said. "Don't you ever wonder about some of the things you get from the beep? That 'cast Dana Lje picked up from Canes Venatici, for instance, the one from the ship that was traveling backward in time? How is that possible? What could be the purpose? Is it——"

"*Pace, pace,*" Krasna said. "I don't know and I don't care. Neither should you. That event is too far in the future for us to worry about. We can't possibly know its context yet, so there's no sense in trying to understand it. If an Englishman of around 1600 had found out about the American Revolution, he would have thought it a tragedy; an Englishman of 1950 would have a very different view of it. We're in the same spot. The messages we get from the really far future have no contexts as yet."

"I think I see," Jo said. "I'll get used to it in time, I suppose, after I use the Dirac for a while. Or does my new rank authorize me to do that?"

"Yes, it does. But, Jo, first I want to pass on to you a rule of Service etiquette that must never be broken. You won't be allowed anywhere near a Dirac mike until you have it burned into your memory beyond any forgetfulness."

"I'm listening, Kras, believe me."

"Good. This is the rule: *The date of a Serviceman's death must never be mentioned in a Dirac 'cast.*"

Jo blinked, feeling a little chilly. The reason behind the rule was decidedly tough-minded, but its ultimate kindness was plain. He said, "I won't forget that. I'll want that protection myself. Many thanks, Kras. What's my new assignment?"

"To begin with," Krasna said, grinning, "as simple a job as I've ever given you, right here on Randolph. Skin out of here and find me that cab driver—the one who mentioned time-travel to you. He's uncomfortably close to the truth; closer than you were in one category.

"Find him, and bring him to me. The Service is about to take in a new raw recruit!"

# This Earth of Hours

THE ADVANCE squadron was coming into line as Master Sergeant Oberholzer came onto the bridge of the *Novoe Washingtongrad,* saluted, and stood stiffly to the left of Lieutenant Campion, the exec, to wait for orders. The bridge was crowded and crackling with tension, but after twenty years in the Marines it was all old stuff to Oberholzer. The *Hobo* (as most of the enlisted men called her, out of earshot of the brass) was at the point of the formation, as befitted a virtually indestructible battleship already surfeited with these petty conquests. The rest of the cone was sweeping on ahead, in the swift enveloping maneuver which had reduced so many previous planets before they had been able to understand what was happening to them.

This time, the planet at the focus of all those shifting conic sections of raw naval power was a place called Callë. It was showing now on a screen that Oberholzer could see, turning as placidly as any planet turned when you were too far away from it to see what guns it might be pointing at you. Lieutenant Campion was watching it too, though he had to look out of the very corners of his eyes to see it at all.

If the exec were caught watching the screen instead of the meter board assigned to him, Captain Hammer would probably reduce him to an ensign. Nevertheless, Campion never took his eyes off the image of Callë. This one was going to be rough.

Captain Hammer was watching, too. After a moment he said, "Sound!" in a voice like sandpaper.

"By the pulse six, sir," Lieutenant Spring's voice murmured from the direction of the 'scope. His junior, a very raw youngster named Rover, passed him a chit from the plotting table. "For that read: By the briefs five eight nine, sir," the invisible navigator corrected.

Oberholzer listened without moving while Captain Hammer muttered under his breath to Flo-Mar 12-Upjohn, the only civilian allowed on the bridge—and small wonder, since he was the Consort of State of the Matriarchy itself. Hammer had long ago become accustomed enough to his own bridge to be able to control who overheard him, but

12-Upjohn's answering whisper must have been audible to every man there.

"The briefing said nothing about a second inhabited planet," the Consort said, a little peevishly. "But then there's very little we *do* know about this system—that's part of our trouble. What makes you think it's a colony?"

"A colony from Callë, not one of ours," Hammer said, in more or less normal tones; evidently he had decided against trying to keep only half of the discussion private. "The electromagnetic 'noise' from both planets has the same spectrum—the energy level, the output, is higher on Callë, that's all. That means similar machines being used in similar ways. And let me point out, Your Excellency, that the outer planet is in opposition to Callë now, which will put it precisely in our rear if we complete this maneuver."

"*When* we complete this maneuver," 12-Upjohn said firmly. "Is there any evidence of communication between the two planets?"

Hammer frowned. "No," he admitted.

"Then we'll regard the colonization hypothesis as unproved—and stand ready to strike back hard if events prove us wrong. I think we have a sufficient force here to reduce *three* planets like Callë if we're driven to that pitch."

Hammer grunted and resigned the argument. Of course it was quite possible that 12-Upjohn was right; he did not lack for experience—in fact, he wore the Silver Earring, as the most-traveled Consort of State ever to ride the Standing Wave. Nevertheless Oberholzer repressed a sniff with difficulty. Like all the military, he was a colonial; he had never seen the Earth, and never expected to; and, both as a colonial and as a Marine who had been fighting the Matriarchy's battles all his adult life, he was more than a little contemptuous of Earthmen, with their tandem names and all that they implied. Of course it was not the Consort of State's fault that he had been born on Earth, and so had been named only Marvin 12 out of the misfortune of being a male; nor that he had married into Florence Upjohn's cabinet, that being the only way one could become a cabinet member, and Marvin 12 having been taught from birth to believe such a post the highest honor a man might covet. All the same, neither 12-Upjohn nor his entourage of drones filled Oberholzer with confidence.

Nobody, however, had asked M. Sgt. Richard Oberholzer what he thought, and nobody was likely to. As the chief of all the non-Navy enlisted personnel on board the *Hobo*, he was expected to be on the bridge when matters were ripening toward criticality; but his duty there was to listen, not to proffer advice. He could not in fact remember any occasion when an officer had asked his opinion, though he had received—and executed—his fair share of near-suicidal orders from bridges long demolished.

"By the pulse five point five," Lieutenant Spring's voice sang.

"Sergeant Oberholzer," Hammer said.

"Aye, sir."

"We are proceeding as per orders. You may now brief your men and put them into full battle gear."

Oberholzer saluted and went below. There was little enough he could tell the squad—as 12-Upjohn had said, Callë's system was nearly unknown—but even that little would improve the total ignorance in which they had been kept till now. Luckily, they were not much given to asking questions of a strategic sort; like impressed spacehands everywhere, the huge mass of the Matriarchy's interstellar holdings meant nothing to them but endlessly riding the Standing Wave, with battle and death lurking at the end of every jump. Luckily also, they were inclined to trust Oberholzer, if only for the low cunning he had shown in keeping most of them alive, especially in the face of unusually Crimean orders from the bridge.

This time Oberholzer would need every ounce of trust and erg of obedience they would give him. Though he never expected anything but the worst, he had a queer cold feeling that this time he was going to get it. There were hardly any data to go on yet, but there had been something about Callë that looked persuasively like the end of the line.

Very few of the forty men in the wardroom even looked up as Oberholzer entered. They were checking their gear in the dismal light of the fluorescents, with the single-mindedness of men to whom a properly wound gun-tube coil, a properly set face-shield gasket, a properly fueled and focused vaulting jet, have come to mean more than parents, children, retirement pensions, the rule of law, or the logic of empire. The only man to show any flicker of interest was Sergeant Cassirir—as was normal, since he was Oberholzer's understudy—and he did no more than look up from over the straps of his antigas suit and say, "Well?"

"Well," Oberholzer said, "now hear this."

There was a sort of composite jingle and clank as the men lowered their gear to the deck or put it aside on their bunks.

"We're investing a planet called Callë in the Canes Venatici cluster," Oberholzer said, sitting down on an olive-drab canvas pack stuffed with lysurgic acid grenades. "A cruiser called the *Assam Dragon*—you were with her on her shakedown, weren't you, Himber?—touched down here ten years ago with a flock of tenders and got swallowed up. They got two or three quick yells for help out and that was that—nothing anybody could make much sense of, no weapons named or description of the enemy. So here we are, loaded for the kill."

"Wasn't any Calley in command of the *Assam Dragon* when I was aboard," Himber said doubtfully.

"Nah. Place was named for the astronomer who spotted her, from the rim of the cluster, a hundred years ago," Oberholzer said. "Nobody names planets for ship captains. Anybody got any sensible questions?"

"Just what kind of trouble are we looking for?" Cassirir said.

"That's just it —we don't know. This is closer to the center of the Galaxy than we've ever gotten before. It may be a population center too; could be that Callë is just one piece of a federation, at least inside its own cluster. That's why we've got the boys from Momma on board; this one could be damn important."

Somebody sniffed. "If this cluster is full of people, how come we never picked up signals from it?"

"How do you know we never did?" Oberholzer retorted. "For all I know, maybe that's why the *Assam Dragon* came here in the first place. Anyhow that's not our problem. All we're——"

The lights went out. Simultaneously, the whole mass of the *Novoe Washingtongrad* shuddered savagely, as though a boulder almost as big as she was had been dropped on her.

Seconds later, the gravity went out too.

## 2

Flo-Mar 12-Upjohn knew no more of the real nature of the disaster than did the wardroom squad, nor did anybody on the bridge, for that matter. The blow had been inde-tectable until it struck, and then most of the fleet was simply annihilated; only the *Hobo* was big enough to survive the blow, and she survived only partially—in fact, in five pieces. Nor did the Consort of State ever know by what miracle the section he was in hit Callë still partially under power; he was not privy to the self-salvaging engineering principles of battleships. All he knew—once he struggled back to consciousness—was that he was still alive, and that there was a broad shaft of sunlight coming through a top-to-bottom split in one wall of what had been his office aboard ship.

He held his ringing head for a while, then got up in search of water. Nothing came out of the dispenser, so he unstrapped his dispatch case from the underside of his desk and produced a pint palladium flask of vodka. He had screwed up his face to sample this—at the moment he would have preferred water—when a groan reminded him that there might be more than one room in his suddenly shrunken universe, as well as other survivors.

He was right on both counts. Though the ship section he was in consisted mostly of engines of whose function he had no notion, there were also three other staterooms. Two of these were deserted, but the third turned out to contain a battered member of his own staff, by name Robin One.

The young man was not yet conscious and 12-Upjohn regarded him with a faint touch of despair. Robin One was perhaps the last man in space that the Consort of State would have chosen to be shipwrecked with.

That he was utterly expendable almost went without saying; he was, after all, a drone. When the perfection of sperm electrophoresis had enabled parents for the first time to predetermine the sex of their children, the predictable result had been an enormous glut of males—which was directly accountable for the present regime on Earth. By the time the people and the lawmakers, thoroughly frightened by the crazy years of fashion upheavals, "beefcake," polyandry, male prostitution, and all the rest, had come to their senses, the Matriarchy was in to stay; a weak electric current had overturned civilized society as drastically as the steel knife had demoralized the Eskimos.

Though the tide of excess males had since receded somewhat, it had left behind a wrack, of which Robin One was a bubble. He was a drone, and hence superfluous by definition—fit only to be sent colonizing, on diplomatic missions or otherwise thrown away.

Superfluity alone, of course, could hardly account for his

presence on 12-Upjohn's staff. Officially, Robin One was an interpreter; actually—since nobody could know the language the Consort of State might be called upon to understand on this mission—he was a poet, a class of unattached males with special privileges in the Matriarchy, particularly if what they wrote was of the middling-difficult or Hillyer Society sort. Robin One was an eminently typical member of this class, distractible, sulky, jealous, easily wounded, homosexual, lazy except when writing, and probably (to give him the benefit of the doubt, for 12-Upjohn had no ear whatever for poetry) the second-worst poet of his generation.

It had to be admitted that assigning 12-Upjohn a poet as an interpreter on this mission had not been a wholly bad idea, and that if Hildegard Muller of the Interstellar Understanding Commission had not thought of it, no mere male would have been likely to—least of all Bar-Rob 4-Agberg, Director of Assimilation. The nightmare of finding the whole of the center of the Galaxy organized into one vast federation, much older than Earth's, had been troubling the State Department for a long time, at first from purely theoretical considerations—all those heart-stars were much older than those in the spiral arms, and besides, where star density in space is so much higher, interstellar travel does not look like quite so insuperable an obstacle as it long had to Earthmen —and later from certain practical signs, of which the obliteration of the *Assam Dragon* and her tenders had been only the most provocative. Getting along with these people on the first contact would be vital, and yet the language barrier might well provoke a tragedy wanted by neither side, as the obliteration of Nagasaki in World War II had been provoked by the mistranslation of a single word. Under such circumstances, a man with a feeling for strange words in odd relationships might well prove to be useful, or even vital.

Nevertheless, it was with a certain grim enjoyment that 12-Upjohn poured into Robin One a good two-ounce jolt of vodka. Robin coughed convulsively and sat up, blinking.

"Your Excellency—how—what's happened? I thought we were dead. But we've got lights again, and gravity."

He was observant, that had to be granted. "The lights are ours but the gravity is Callë's," 12-Upjohn explained tersely. "We're in a part of the ship that cracked up."

"Well, it's good that we've got power."

"We can't afford to be philosophical about it. Whatever

shape it's in, this derelict is a thoroughly conspicuous object and we'd better get out of it in a hurry."

"Why?" Robin said. "We were supposed to make contact with these people. Why not just sit here until they notice and come to see us?"

"Suppose they just blast us to smaller bits instead? They didn't stop to parley with the fleet, you'll notice."

"This is a different situation," Robin said stubbornly. "I wouldn't have stopped to parley with that fleet myself, if I'd had the means of knocking it out first. It didn't look a bit like a diplomatic mission. But why should they be afraid of a piece of a wreck?"

The Consort of State stroked the back of his neck reflectively. The boy had a point. It was risky; on the other hand, how long would they survive foraging in completely unknown territory? And yet obviously they couldn't stay cooped up in here forever—especially if it was true that there was already no water.

He was spared having to make up his mind by a halloo from the direction of the office. After a startled stare at each other, the two hit the deck running.

Sergeant Oberholzer's face was peering grimly through the split in the bulkhead.

"Oho," he said. "So you *did* make it." He said something unintelligible to some invisible person outside, and then squirmed through the breach into the room, with considerable difficulty, since he was in full battle gear. "None of the officers did, so I guess that puts you in command."

"In command of what?" 12-Upjohn said dryly.

"Not very much," the Marine admitted. "I've got five men surviving, one of them with a broken hip, and a section of the ship with two drive units in it. It would lift, more or less, if we could jury-rig some controls, but I don't know where we'd go in it without supplies or a navigator—or an overdrive, for that matter." He looked about speculatively. "There was a Standing Wave transceiver in this section, I think, but it'd be a miracle if it still functioned."

"Would you know how to test it?" Robin asked.

"No. Anyhow we've got more immediate business than that. We've picked up a native. What's more, he speaks English—must have picked it up from the *Assam Dragon*. We started to ask him questions, but it turns out he's some sort of top official, so we brought him over here on the off chance that one of you was alive."

"What a break!" Robin One said explosively.

"A whole series of them," 12-Upjohn agreed, none too happily. He had long ago learned to be at his most suspicious when the breaks seemed to be coming his way. "Well, better bring him in."

"Can't," Oberholzer said. "Apologies, Your Excellency, but he wouldn't fit. You'll have to come to him."

## 3

It was impossible to imagine what sort of stock the Callëan had evolved from. He seemed to be a thoroughgoing mixture of several different phyla. Most of him was a brown, segmented tube about the diameter of a barrel and perhaps twenty-five feet long, rather like a cross between a python and a worm. The front segments were carried upright, raising the head a good ten feet off the ground.

Properly speaking, 12-Upjohn thought, the Callëan really had no head, but only a front end, marked by two enormous faceted eyes and three upsetting simple eyes which were usually closed. Beneath these there was a collar of six short, squidlike tentacles, carried wrapped around the creature in a ropy ring. He was as impossible-looking as he was fearsome, and 12-Upjohn felt at a multiple disadvantage from the beginning.

"How did you learn our language?" he said, purely as a starter.

"I learned it from you," the Callëan said promptly. The voice was unexpectedly high, a quality which was accentuated by the creature's singsong intonation; 12-Upjohn could not see where it was coming from. "From your ship which I took apart, the dragon-of-war."

"Why did you do that?"

"It was evident that you meant me ill," the Callëan sang. "At that time I did not know that you were sick, but that became evident at the dissections."

"Dissections! You dissected the crew of the *Dragon?*"

"All but one."

There was a growl from Oberholzer. The Consort of State shot him a warning glance.

"You may have made a mistake," 12-Upjohn said. "A natural mistake, perhaps. But it was our purpose to offer you trade and peaceful relationships. Our weapons were only precautionary."

"I do not think so," the Callëan said, "and I never make

mistakes. That you make mistakes is natural, but it is not natural to me."

12-Upjohn felt his jaw dropping. That the creature meant what he said could not be doubted; his command of the language was too complete to permit any more sensible interpretation. 12-Upjohn found himself at a loss; not only was the statement the most staggering he had ever heard from any sentient being, but while it was being made he had discovered how the Callëan spoke: the sounds issued at low volume from a multitude of spiracles or breath-holes all along the body, each hole producing only one pure tone, the words and intonations being formed in mid-air by inter-modulation—a miracle of co-ordination among a multitude of organs obviously unsuitable for sound-forming at all. This thing was formidable—that would have been evident even without the lesson of the chunk of the *Novoe Washington-grad* canted crazily in the sands behind them.

Sands? He looked about with a start. Until that moment the Callëan had so hypnotized his attention that he had for-gotten to look at the landscape, but his unconscious had registered it. Sand, and nothing but sand. If there were better parts of Callë than this desert, they were not visible from here, all the way to the horizon.

"What do you propose to do with us?" he said at last. There was really nothing else to say; cut off in every possible sense from his home world, he no longer had any base from which to negotiate.

"Nothing," the Callëan said. "You are free to come and go as you please."

"You're no longer afraid of us?"

"No. When you came to kill me I prevented you, but you can no longer do that."

"There you've made a mistake, all right," Oberholzer said, lifting his rifle toward the multicolored, glittering jewels of the Callëan's eyes. "You know what this is—they must have had them on the *Dragon*."

"Don't be an idiot, Sergeant," 12-Upjohn said sharply. "We're in no position to make any threats." Nor, he added silently, should the Marine have called attention to his gun before the Callëan had taken any overt notice of it.

"I know what it is," the creature said. "You cannot kill me with that. You tried it often before and found you could not. You would remember this if you were not sick."

"I never saw anything that I couldn't kill with a Sussmann

flamer," Oberholzer said between his teeth. "Let me try it on the bastard, Your Excellency."

"Wait a minute," Robin One said, to 12-Upjohn's astonishment. "I want to ask some questions—if you don't mind, Your Excellency?"

"I don't mind," 12-Upjohn said after an instant. Anything to get the Marine's crazy impulse toward slaughter sidetracked. "Go ahead."

"Did you dissect the crew of the *Assam Dragon* personally?" Robin asked the Callëan.

"Of course."

"Are you the ruler of this planet?"

"Yes."

"Are you the only person in this system?"

"No."

Robin paused and frowned. Then he said: "Are you the only person of your species in your system?"

"No. There is another on Xixobrax—the fourth planet."

Robin paused once more, but not, it seemed to 12-Upjohn, as though he were in any doubt; it was only as though he were gathering his courage for the key question of all. 12-Upjohn tried to imagine what it might be, and failed.

"How many of you are there?" Robin One said.

"I cannot answer that. As of the instant you asked me that question, there were eighty-three hundred thousand billion, one hundred and eighty nine million, four hundred and sixty five thousand, one hundred and eighty; but now the number has changed, and it goes on changing."

"Impossible," 12-Upjohn said, stunned. "Not even two planets could support such a number—and you'd never allow a desert like this to go on existing if you had even a fraction of that population to support. I begin to think, sir, that you are a type normal to my business: the ordinary, unimaginative liar."

"He's not lying," Robin said, his voice quivering. "It all fits together. Just let me finish, sir, please. I'll explain, but I've got to go through to the end first."

"Well," 12-Upjohn said, helplessly, "all right, go ahead." But he was instantly sorry, for what Robin One said was:

"Thank you. I have no more questions."

The Callëan turned in a great liquid wheel and poured away across the sand dunes at an incredible speed. 12-Upjohn shouted after him, without any clear idea of what it was that he was shouting—but no matter, for the Callëan took no notice. Within seconds, it seemed, he was only a thread-worm in the middle distance, and then he was gone. They were all alone in the chill desert air.

Oberholzer lowered his rifle bewilderedly. "He's fast," he said to nobody in particular. "Cripes, but he's fast. I couldn't even keep him in the sights."

"That proves it," Robin said tightly. He was trembling, but whether with fright or elation, 12-Upjohn could not tell; possibly both.

"It had better prove something," the Consort of State said, trying hard not to sound portentous. There was something about this bright remote desert that made empty any possible pretense to dignity. "As far as I can see, you've just lost us what may have been our only chance to treat with these creatures . . . just as surely as the sergeant would have done it with his gun. Explain, please."

"I didn't really catch on until I realized that he was using the second person singular when he spoke to us," Robin said. If he had heard any threat implied in 12-Upjohn's charge, it was not visible; he seemed totally preoccupied. "There's no way to tell them apart in modern English. We thought he was referring to us as 'you' plural, but he wasn't, any more than his 'I' was a plural. He thinks we're all a part of the same personality—including the men from the *Dragon*, too—*just as he is himself*. That's why he left when I said I had no more questions. He can't comprehend that each of us has an independent ego. For him such a thing doesn't exist."

"Like ants?" 12-Upjohn said slowly. "I don't see how an advanced technology . . . but no, I do see. And if it's so, it means that any Calléan we run across could be their chief of state, but that no one of them actually is. The only other real individual is next door, on the fourth planet—another hive ego."

"Maybe not," Robin said. "Don't forget that he thinks we're part of one, too."

12-Upjohn dismissed that possibility at once. "He's sure to know his own system, after all. . . . What alarms me is the population figure he cited. It's got to be *at least* clusterwide— and from the exactness with which he was willing to cite it, for a given instant, he had to have immediate access to it. An instant, effortless census."

"Yes," Robin said. "Meaning mind-to-mind contact, from

one to all, throughout the whole complex. That's what started me thinking about the funny way he used pronouns."

"If that's the case, Robin, we are *spurlos versenkt*. And my pronoun includes the Earth."

"They may have some limitations," Robin said, but it was clear that he was only whistling in the dark. "But at least it explains why they butchered the *Dragon's* crew so readily —and why they're willing to let us wander around their planet as if we didn't even exist. We don't, for them. They can't have any respect for a single life. No wonder they didn't give a damn for the sergeant's gun!"

His initial flush had given way to a marble paleness; there were beads of sweat on his brow in the dry hot air, and he was trembling harder than ever. He looked as though he might faint in the next instant, though only the slightest of stutters disturbed his rush of words. But for once the Consort of State could not accuse him of agitation over trifles.

Oberholzer looked from one to the other, his expression betraying perhaps only disgust, or perhaps blank incomprehension—it was impossible to tell. Then, with a sudden sharp *snick* which made them both start, he shot closed the safety catch on the Sussmann.

"Well," he said in a smooth cold empty voice, "now we know what we'll eat."

## 4

Their basic and dangerous division of plans and purposes began with that.

Sergeant Oberholzer was not a fool, as the hash marks on his sleeve and the battle stars on his ribbons attested plainly; he understood the implications of what the Callëan had said—at least after the Momma's boy had interpreted them; and he was shrewd enough not to undervalue the contribution the poor terrified fairy had made to their possible survival on this world. For the moment, however, it suited the Marine to play the role of the dumb sergeant to the hilt. If a full understanding of what the Callëans were like might reduce him to a like state of trembling impotence, he could do without it.

Not that he really believed that any such thing could happen to him; but it was not hard to see that Momma's boys were halfway there already—and if the party as a whole hoped to get anything done, they had to be jolted out of it as fast as possible.

At first he thought he had made it. "Certainly not!" the Consort of State said indignantly. "You're a man, sergeant, not a Calléan. Nothing the Calléans do is any excuse for your behaving otherwise than as a man."

"I'd rather eat an enemy than a friend," Oberholzer said cryptically. "Have you got any supplies inside there?"

"I—I don't know. But that has nothing to do with it."

"Depends on what you mean by 'it.' But maybe we can argue about that later. What are your orders, Your Excellency?"

"I haven't an order in my head," 12-Upjohn said with sudden, disarming frankness. "We'd better try to make some sensible plans first, and stop bickering. Robin, stop snuffling, too. The question is, what can we do besides trying to survive, and cherishing an idiot hope for a rescue mission?"

"For one thing, we can try to spring the man from the *Dragon's* crew that these worms have still got alive," Oberholzer said. "If that's what he meant when he said they dissected all but one."

"That doesn't seem very feasible to me," 12-Upjohn said. "We have no idea where they're holding him——"

"Ask them. This one answered every question you asked him."

"—and even supposing that he's near by, we couldn't free him from a horde of Calléans, no matter how many dead bodies they let you pile up. At best, sooner or later you'd run out of ammunition."

"It's worth trying," Oberholzer said. "We could use the manpower."

"What for?" Robin One demanded. "He'd be just one more mouth to feed. At the moment, at least, they're feeding him."

"For raising ship," Oberholzer retorted. "*If* there's any damn chance of welding our two heaps of junk together and getting off this mudball. We ought to look into it, anyhow."

Robin One was looking more alarmed by the minute. If the prospect of getting into a fight with the Calléans had scared him, Oberholzer thought, the notion of hard physical labor evidently was producing something close to panic.

"Where could we go?" he said. "Supposing that we could fly such a shambles at all?"

"I don't know," Oberholzer said. "We don't know what's possible yet. But anything's better than sitting around here and starving. First off, I want that man from the *Dragon*."

"I'm opposed to it," 12-Upjohn said firmly. "The Callëans are leaving us to our own devices now. If we cause any real trouble they may well decide that we'd be safer locked up, or dead. I don't mind planning to lift ship if we can—but no military expeditions."

"Sir," Oberholzer said, "military action on this planet is what I was sent here for. I reserve the right to use my own judgment. You can complain, if we ever get back—but I'm not going to let a man rot in a worm-burrow while I've got a gun on my back. You can come along or not, but we're going."

He signaled to Cassirir, who seemed to be grinning slightly. 12-Upjohn stared at him for a moment, and then shook his head.

"We'll stay," he said. "Since we have no water, Sergeant, I hope you'll do us the kindness of telling us where your part of the ship lies."

"That way, about two kilometers," Oberholzer said. "Help yourself. If you want to settle in there, you'll save us the trouble of toting Private Hannes with us on a stretcher."

"Of course," the Consort of State said. "We'll take care of him. But, Sergeant . . ."

"Yes, Your Excellency?"

"If this stunt of yours still leaves us all alive afterwards, and we do get back to any base of ours, I will *certainly* see to it that a complaint is lodged. I'm not disowning you now because it's obvious that we'll all have to work together to survive, and a certain amount of amity will be essential. But don't be deceived by that."

"I understand, sir," Oberholzer said levelly. "Cassirir, let's go. We'll backtrack to where we nabbed the worm, and then follow his trail to wherever he came from. Fall in."

The men shouldered their Sussmanns. 12-Upjohn and Robin One watched them go. At the last dune before the two would go out of sight altogether, Oberholzer turned and waved, but neither waved back. Shrugging, Oberholzer resumed plodding.

"Sarge?"

"Yeah?"

"How *do* you figure to spring this joker with only four guns?"

"Five guns if we spring him—I've got a side arm," Oberholzer reminded him. "We'll play it by ear, that's all. I want to see just how serious these worms are about leaving us alone, and letting us shoot them if we feel like it. I've got a hunch that they aren't very bright, one at a time, and don't react fast to strictly local situations. If this whole planet is like one huge body, and the worms are its brain cells, then we're germs—and maybe it'd take more than four germs to make the body do anything against us that counted, at least fast enough to do any good."

Cassirir was frowning absurdly; he did not seem to be taking the theory in without pain. Well, Cassirir had never been much of a man for tactics.

"Here's where we found the guy," one of the men said, pointing at the sand.

"That's not much of a trail," Cassirir said. "If there's any wind it'll be wiped out like a shot."

"Take a sight on it, that's all we need. You saw him run off—straight as a ruled line, no twists or turns around the dunes or anything. Like an army ant. If the trail sands over, we'll follow the sight. It's a cinch it leads someplace."

"All right," Cassirir said, getting out his compass. After a while the four of them resumed trudging.

There were only a few drops of hot, flat-tasting water left in the canteens, and their eyes were gritty and red from dryness and sand, when they topped the ridge that overlooked the nest. The word sprang instantly into Oberholzer's mind, though perhaps he had been expecting some such thing ever since Robin One had compared the Calléans to ants.

It was a collection of rough white spires, each perhaps fifty feet high, rising from a common doughlike mass which almost filled a small valley. There was no greenery around it and no visible source of water, but there were three roads, two of them leading into oval black entrances which Oberholzer could see from here. Occasionally—not often—a Calléan would scuttle out and vanish, or come speeding over the horizon and dart into the darkness. Some of the spires bore masts carrying what seemed to be antennae or more recondite electronic devices, but there were no windows to be seen; and the only sound in the valley, except for the dry dusty wind, was a subdued composite hum.

"Man!" Cassirir said, whispering without being aware of it. "It must be as black as the ace of spades in there. Anybody got a torch?"

Nobody had. "We won't need one anyhow," Oberholzer

said confidently. "They've got eyes, and they can see in desert sunlight. That means they can't move around in total darkness. Let's go—I'm thirsty."

They stumbled down into the valley and approached the nearest black hole cautiously. Sure enough, it was not as black as it had appeared from the hill; there was a glow inside, which had been hidden from them against the contrast of the glaringly lit sands. Nevertheless, Oberholzer found himself hanging back.

While he hesitated, a Callëan came rocketing out of the entrance and pulled to a smooth, sudden stop.

"You are not to get in the way," he said, in exactly the same piping singsong voice the other had used.

"Tell me where to go and I'll stay out of your way," Oberholzer said. "Where is the man from the warship that you didn't dissect?"

"In Gnitonis, halfway around the world from here."

Oberholzer felt his shoulders sag, but the Callëan was not through. "You should have told me that you wanted him," he said. "I will have him brought to you. Is there else that you need?"

"Water," Oberholzer said hopefully.

"That will be brought. There is no water you can use here. Stay out of the cities; you will be in the way."

"How else can we eat?"

"Food will be brought. You should make your needs known; you are of low intelligence and helpless. I forbid nothing, I know you are harmless, and your life is short in any case; but I do not want you to get in the way."

The repetition was beginning to tell on Oberholzer, and the frustration created by his having tried to use a battering ram against a freely swinging door was compounded by his mental picture of what the two Momma's boys would say when the squad got back.

"Thank you," he said, and bringing the Sussmann into line, he trained it on the Callëan's squidlike head and squeezed the trigger.

It was at once established that the Callëans were as mortal to Sussmann flamers as is all other flesh and blood; this one made a very satisfactory corpse. Unsatisfied, the flamer bolt went on to burn a long slash in the wall of the nest, not far above the entrance. Oberholzer grounded the rifle and waited to see what would happen next; his men hefted their weapons tensely.

For a few minutes there was no motion but the random twitching of the headless Callëan's legs. Evidently he was still not entirely dead, though he was a good four feet shorter than he had been before, and plainly was feeling the lack. Then, there was a stir inside the dark entrance.

A ten-legged animal about the size of a large rabbit emerged tentatively into the sunlight, followed by two more, and then by a whole series of them, perhaps as many as twenty. Though Oberholzer had been unabashed by the Callëans themselves, there was something about these things that made him feel sick. They were coal black and shiny, and they did not seem to have any eyes; their heavily armored heads bore nothing but a set of rudimentary palps and a pair of enormous pincers, like those of a June beetle.

Sightless or no, they were excellent surgeons. They cut the remains of the Callëan swiftly into sections, precisely one metamere to a section, and bore the carrion back inside the nest. Filled with loathing, Oberholzer stepped quickly forward and kicked one of the last in the procession. It toppled over like an unstable kitchen stool, but regained its footing as though nothing had happened. The kick had not hurt it visibly, though Oberholzer's toes felt as though he had kicked a Victorian iron dog. The creature, still holding its steak delicately in its living tongs, mushed implacably after the others back into the dubiety of the nest. Then all that was left in the broiling sunlight was a few pools of blackening blood seeping swiftly into the sand.

"Let's get out of here," Cassirir said raggedly.

"Stand fast," Oberholzer growled. "If they're mad at us, I want to know about it right now."

But the next Callëan to pass them, some twenty eternal minutes later, hardly even slowed down. "Keep out of the way," he said, and streaked away over the dunes. Snarling, Oberholzer caromed a bolt after him, but missed him clean.

"All right," he said. "Let's go back. No hitting the canteens till we're five kilometers past the mid-point cairn. March!"

The men were all on the verge of prostration by the time that point was passed, but Oberholzer never once had to enforce the order. Nobody, it appeared, was eager to come to an end on Callë as a series of butcher's cuts in the tongs of a squad of huge black beetles.

## 5

"I know what they think," the man from the *Assam Dragon* said. "I've heard them say it often enough."

He was a personable youngster, perhaps thirty, with blond wavy hair which had been turned almost white by the strong Callëan sunlight: his captors had walked him for three hours every day on the desert. He had once been the *Assam Dragon's* radioman, a post which in interstellar flight is a branch of astronomy, not of communications; nevertheless, Oberholzer and the marines called him Sparks, in deference to a tradition which, 12-Upjohn suspected, the marines did not even know existed.

"Then why wouldn't there be a chance of our establishing better relations with the 'person' on the fourth planet?" 12-Upjohn said. "After all, there's never been an Earth landing there."

"Because the 'person' on Xixobrax is a colony of Callë, and knows everything that goes on here. It took the two planets in co-operation to destroy the fleet. There's almost full telepathic communion between the two—in fact, all through the Central Empire. The only rapport that seems to weaken over short distances—interplanetary distances—is the sense of identity. That's why each planet has an 'I' of its own, its own ego. But it's not the kind of ego we know anything about. Xixobrax wouldn't give us any better deal than Callë has, any more than I'd give Callë a better deal than you would, Your Excellency. They have common purposes and allegiances. All the Central Empire seems to be like that."

12-Upjohn thought about it; but he did not like what he thought. It was a knotty problem, even in theory.

Telepathy among men had never amounted to anything. After the pioneer exploration of the microcosm with the Arpe Effect—the second of two unsuccessful attempts at an interstellar drive, long before the discovery of the Standing Wave—it had become easy to see why this would be so. Psi forces in general were characteristic only of the subspace in which the primary particles of the atom had their being; their occasional manifestations in the macrocosm were statistical accidents, as weak and indirigible as spontaneous radioactive decay.

Up to now this had suited 12-Upjohn. It had always seemed to him that the whole notion of telepathy was a

dodge—an attempt to by-pass the plain duty of each man
to learn to know his brother, and, if possible, to learn to
love him; the telepathy fanatics were out to short-circuit the
task, to make easy the most difficult assignment a human
being might undertake. He was well aware, too, of the bias
against telepathy which was inherent in his profession of
diplomat; yet he had always been certain of his case, hazy
though it was around the edges. One of his proofs was that
telepathy's main defenders invariably were incorrigibly lazy
writers, from Upton Sinclair and Theodore Dreiser all the
way down to . . .

All the same, it seemed inarguable that the whole center
of the Galaxy, an enormously diverse collection of peoples
and cultures, was being held together in a common and
strife-free union by telepathy alone, or perhaps by telepathy
and its even more dubious adjuncts: a whole galaxy held
together by a force so unreliable that two human beings
sitting across from each other at a card table had never
been able to put it to an even vaguely practicable use.

Somewhere, there was a huge hole in the argument.

While he had sat helplessly thinking in these circles, even
Robin One was busy, toting power packs to the welding crew
which was working outside to braze together on the desert the
implausible, misshapen lump of metal which the Marine
sergeant was fanatically determined would become a ship
again. Now the job was done, though no shipwright would
admire it, and the question of where to go with it was being
debated in full council. Sparks, for his part, was prepared
to bet that the Calléans would not hinder their departure.

"Why would they have given us all this oxygen and stuff
if they were going to prevent us from using it?" he said
reasonably. "They know what it's for—even if they have
no brains, collectively they're plenty smart enough."

"*No* brains?" 12-Upjohn said. "Or are you just exag-
gerating?"

"No brains," the man from the *Assam Dragon* insisted.
"Just lots of ganglia. I gather that's the way all of the races
of the Central Empire are organized, regardless of other
physical differences. That's what they mean when they say
we're all sick—hadn't you realized that?"

"No," 12-Upjohn said in slowly dawning horror. "You
had better spell it out."

"Why, they say that's why we get cancer. They say that

the brain is the ultimate source of all tumors, and is itself a tumor. They call it 'hostile symbiosis.' "

"Malignant?"

"In the long run. Races that develop them kill themselves off. Something to do with solar radiation; animals on planets of Population II stars develop them, Population I planets don't."

Robin One hummed an archaic twelve-tone series under his breath. There were no words to go with it, but the Consort of State recognized it; it was part of a chorale from a twentieth-century American opera, and the words went: *Weep, weep beyond time for this Earth of hours.*

"If fits," he said heavily. "So to receive and use a weak field like telepathy, you need a weak brain. Human beings will never make it."

"Earthworms of the galaxy, unite," Robin One said.

"They already have," Sergeant Oberholzer pointed out. "So where does all this leave us?"

"It means," 12-Upjohn said slowly, "that this Central Empire, where the stars are almost all Population I, is spreading out toward the spiral arms where the Earth lies. Any cluster civilizations they meet are natural allies—clusters are purely Population I—and probably have already been mentally assimilated. Any possible natural allies *we* meet, going around Population II stars, we may well pick a fight with instead."

"That's not what I meant," Sergeant Oberholzer said.

"I know what you meant; but this changes things. As I understand it, we have a chance of making a straight hop to the nearest Earth base, if we go on starvation rations——"

"—and if I don't make more than a point zero five per cent error in plotting the course," Sparks put in.

"Yes. On the other hand, we can make *sure* of getting there by going in short leaps via planets known to be inhabited, but never colonized and possibly hostile. The only other possibility is Xixobrax, which I think we've ruled out. Correct?"

"Right as rain," Sergeant Oberholzer said. "Now I see what you're driving at, Your Excellency. The only thing is— you didn't mention that the stepping stone method will take us the rest of our lives."

"So I didn't," 12-Upjohn said bleakly. "But I hadn't forgotten it. The other side of *that* coin is that it will be even longer than that before the Matriarchy and the Central Empire collide."

"After which," Sergeant Oberholzer said with a certain relish, "I doubt that it'll be a Matriarchy, whichever wins. Are you calling for a vote, sir?"

"Well—yes, I seem to be."

"Then let's grasshopper," Sergeant Oberholzer said unhesitatingly. "The boys and I can't fight a point zero five per cent error in navigation—but for hostile planets, we've got the flamers."

Robin One shuddered. "I don't mind the fighting part," he said unexpectedly. "But I *do* simply loathe the thought of being an old, old man when I get home. All the same, we do have to get the word back."

"You're agreeing with the sergeant?"

"Yes, that's what I said."

"I agree," Sparks said. "Either way we may not make it, but the odds are in favor of doing it the hard way."

"Very good," 12-Upjohn said. He was uncertain of his exact emotion at this moment; perhaps gloomy satisfaction was as close a description as any. "I make it unanimous. Let's get ready."

The sergeant saluted and prepared to leave the cabin; but suddenly he turned back.

"I didn't think very much of either of you, a while back," he said brutally. "But I'll tell you this: there must be something about brains that involves guts, too. I'll back 'em any time against any critter that lets itself be shot like a fish in a barrel—whatever the odds."

The Consort of State was still mulling that speech over as the madman's caricature of an interstellar ship groaned and lifted its lumps and angles from Callë. Who knows, he kept telling himself, who knows, it might even be true.

But he noticed that Robin One was still humming the chorale from *Psyche and Eros*; and ahead the galactic night was as black as death.

The End